TEACHER'S EDITION

AN INQUIRY EARTH SCIENCE PROGRAM

INVESTIGATING CLIMATE AND WEATHER

Michael J. Smith Ph.D.
American Geological Institute

John B. Southard Ph.D.
Massachusetts Institute of Technology

Colin Mably
Curriculum Developer

Developed by the American Geological Institute
Supported by the National Science Foundation and
the American Geological Institute Foundation

Published by
It's About Time Inc., Armonk, NY

It's About Time, Inc.
84 Business Park Drive, Armonk, NY 10504
Phone (914) 273-2233 Fax (914) 273-2227
Toll Free (888) 698-TIME
www.Its-About-Time.com

Publisher
Laurie Kreindler

Project Editor
Ruta Demery

Contributing Writers
William Jones
Matthew Smith

Design
John Nordland

Production Manager
Joan Lee

Associate Editor
Al Mari

All student activities in this textbook have been designed to be as safe as possible, and have been reviewed by professionals specifically for that purpose. As well, appropriate warnings concerning potential safety hazards are included where applicable to particular activities. However, responsibility for safety remains with the student, the classroom teacher, the school principals, and the school board.

Investigating Earth Systems™ is a registered trademark of the American Geological Institute. Registered names and trademarks, etc., used in this publication, even without specific indication thereof, are not to be considered unprotected by law.

It's About Time™ is a registered trademark of It's About Time, Inc. Registered names and trademarks, etc., used in this publication, even without specific indication thereof, are not to be considered unprotected by law.

© Copyright 2002: American Geological Institute

All rights reserved. No part of this publication may be reproduced, stored in a retrieval system, or transmitted, in any form or by any means, electronic, mechanical, photocopying, recording, or otherwise, without the prior written permission of the copyright owner.

Care has been taken to trace the ownership of copyright material contained in this publication. The publisher will gladly receive any information that will rectify any reference or credit line in subsequent editions.

Printed and bound in the United States of America

ISBN #1-58591-091-0

1 2 3 4 5 QC 06 05 04 03 02

This project was supported, in part, by the
National Science Foundation (grant no. 9353035)

Opinions expressed are those of the authors and not necessarily those of the National Science Foundation or the donors of the American Geological Institute Foundation.

Student's Edition Illustrations and Photos

C23, American Meteorological Society;

C10, Annapolis Weathervanes;

C54, illustration by Stuart Armstrong;

C6, C7, C23, C25, C26, C27, C33, C34, C35, C39, C50, C51, C52, C61, C70, C75, C83, C84, C85, illustrations by Burmar Technical Corporation;

C12, Cody Mercantile Catalog;

C65 (left), Corbis Royalty Free Images;

C24, C30, C31, C40, C42, C43, The DataStreme Project, American Meteorological Society;

C11 (2nd down, right), (2nd from bottom right), Digital Royalty Free Images;

C44, Digital Vision Royalty Free Images;

C66 (bottom), Digital Vision Royalty Free Images;

Cv, Cxii, C2, C15, C25, C28, C41, C49, C63, C70, C82, illustrations by Dennis Falcon;

C72, (bottom), Geoff Hargreaves, USGS/National Ice Core Laboratory;

C67, John Karachewski;

C11 (2nd from bottom, left), (top right), James Koermer, Plymouth State College;

C11, (top left) Ralph Kresge, NOAA;

C73 (bottom), Laboratory of Tree-Ring Research, University of Tennessee;

C71, Steven Manchester; courtesy of Oregon Department of Geology and Mineral Industries;

C65, Martin Miller;

C11 (bottom photos),

C56, C66, (top), C69, C72 (top),

C79, C85, Bruce F. Molnia;

C11 (2nd from top, left),

C28, C45, Joe Moran;

C37, NASA;

C22, NOAA;

C1, C8, C14, C20, C48, C57 (top, bottom), C73 (top), C76, C77, PhotoDisc;

C9, C18, Doug Sherman, Geo File Photography;

C54, source, USGS

Taking Full Advantage of Investigating Earth Systems Through Professional Development

Implementing a new curriculum is challenging. That is why It's About Time Publishing has partnered with the American Geological Institute, developers of *Investigating Earth Systems (IES)*, to provide a full range of professional development services. The sessions described below were designed to help you deepen your understanding of the content, pedagogy, and assessment strategies outlined in this Teacher's Edition, and adapt the program to suit the needs of your students and your local and state standards and curriculum frameworks.

Professional Development Services Available

Implementation Workshops
Two- to five-day sessions held at your site that prepare you to implement the inquiry, systems, and community-based approach to learning Earth Science featured in *IES*. These workshops can be tailored to serve the needs of your school district, with chapters selected from the modules based on local or state curricula and framework criteria.

Program Overviews
One- to three-day introductory sessions that provide a complete overview of the content and pedagogy of the *IES* program, as well as hands-on experience with activities from specific chapters. Program overviews are designed in consultation with school districts, counties, and SSI organizations.

Regional New-Teacher Summer Institutes
Two- to five-day sessions that are designed to deepen your Earth Science content knowledge, and to prepare you to teach through inquiry. Guidance is provided in the gathering and use of appropriate materials and resources and specific attention is directed to the assessment of student learning.

Leadership Institutes
Six-day summer sessions conducted by the American Geological Institute that are designed to prepare current users for professional development leadership and mentoring within their districts or as consultants for It's About Time.

Follow-up Workshops
One- to two-day sessions that provide additional Earth Science content and pedagogy support to teachers using the program. These workshops focus on identifying and solving practical issues and challenges to implementing an inquiry-based program.

Mentoring Visits
One-day visits that can be tailored to your specific needs that include class visits, mentoring teachers of the program, and in-service sessions.

Please fill in the form below to receive more information about participating in one of these Professional Development Services. The form can be directly faxed to our Professional Development at 914-273-2227. Our department will contact you to discuss further details and fees.

District/School: _____ Phone: _____

Address: _____

Contact Name: _____ Title: _____

E-mail: _____ Fax: _____

School Enrollment: _____ Number of Students Impacted: _____ Grade Level: _____

Have you purchased the following: ❏ Student Editions   ❏ Teacher Editions   ❏ Kits

Briefly explain how you plan to implement or how you are implementing the program in your school.

_____

_____

Teacher's Edition

Table of Contents

Investigating Earth Systems Team vi
Acknowledgements viii
The American Geological Institute and *Investigating Earth Systems* xi
Developing *Investigating Earth Systems* xii
Investigating Earth Systems Modules xiii
Investigating Earth Systems:
   Correlation to the National Science Education Standards xiv
Using *Investigating Earth Systems* Features in Your Classroom xvi
Using the *Investigating Earth Systems* Web Site xxx
Enhancing Teacher Content Knowledge xxxi
Managing Inquiry in Your *Investigating Earth Systems* Classroom xxxii
Assessing Student Learning in *Investigating Earth Systems* xxxv
Investigating Earth Systems Assessment Tools xxxviii
Reviewing and Reflecting upon Your Teaching xlii
Enhancing Investigating Climate and Weather with GETIT xliii
Investigating Climate and Weather: Introduction 1
Students' Conceptions about Climate and Weather 3
Investigating Climate and Weather: Module Flow 4
Investigating Climate and Weather: Module Objectives 5
National Science Education Content Standards 8
Key NSES Earth Science Standards Addressed in *IES* Climate and Weather 9
Key AAAS Earth Science Benchmarks
   Addressed in *IES* Climate and Weather 9
Materials and Equipment List for Investigating Climate and Weather 10
Pre-assessment 14
Introducing the Earth System 19
Introducing Inquiry Processes 22
Introducing Climate and Weather 24
Why are Climate and Weather Important? 26
Investigation 1: Observing Weather 29
Investigation 2: Comparing Weather Reports 73
Investigation 3: Weather Maps 101
Investigation 4: Weather Radiosondes, Satellites, and Radar 145
Investigation 5: The Causes of Weather 181
Investigation 6: Climates 223
Investigation 7: Exploring Climate Change 255
Investigation 8: Climate Change Today 291
Reflecting 318
Appendices: Alternative End-of-Module Assessment 322
             Assessment Tools 326
             Blackline Masters 336

Investigating Earth Systems Team

Project Staff

Michael J. Smith, Principal Investigator
    Director of Education, American Geological Institute
John B. Southard, Senior Writer
    Professor of Geology Emeritus, Massachusetts Institute of Technology
Matthew Smith, Project Manager
    American Geological Institute
William O. Jones, Contributing Writer
    American Geological Institute
Caitlin N. Callahan, Project Assistant
    American Geological Institute
William S. Houston, Field Test Coordinator
    American Geological Institute
Harvey Rosenbaum, Field Test Evaluator
    Montgomery County School District, Maryland
Fred Finley, Project Evaluator
    University of Minnesota
Lynn Lindow, Pilot Test Evaluator
    University of Minnesota

Original Project Personnel

Robert L. Heller, Principal Investigator
Charles Groat, United States Geological Survey
Colin Mably, LaPlata, Maryland
Robert Ridky, University of Maryland
Marilyn Suiter, American Geological Institute

Teacher's Edition

National Advisory Board

Jane Crowder
    Middle School Teacher, WA
Kerry Davidson
    Louisiana Board of Regents, LA
Joseph D. Exline
    Educational Consultant, VA
Louis A. Fernandez
    California State University, CA
Frank Watt Ireton
    National Earth Science Teachers Association, DC
LeRoy Lee
    Wisconsin Academy of Sciences, Arts and Letters, WI
Donald W. Lewis
    Chevron Corporation, CA
James V. O'Connor (deceased)
    University of the District of Columbia, DC
Roger A. Pielke Sr.
    Colorado State University, CO
Dorothy Stout
    Cypress College, CA
Lois Veath
    Advisory Board Chairperson - Chadron State College, NE

National Science Foundation Program Officers

Gerhard Salinger
Patricia Morse

Acknowledgements

Principal Investigator

Michael Smith is Director of Education at the American Geological Institute in Alexandria, Virginia. Dr. Smith worked as an exploration geologist and hydrogeologist. He began his Earth Science teaching career with Shady Side Academy in Pittsburgh, PA in 1988 and most recently taught Earth Science at the Charter School of Wilmington, DE. He earned a doctorate from the University of Pittsburgh's Cognitive Studies in Education Program and joined the faculty of the University of Delaware School of Education in 1995. Dr. Smith received the Outstanding Earth Science Teacher Award for Pennsylvania from the National Association of Geoscience Teachers in 1991, served as Secretary of the National Earth Science Teachers Association, and is a reviewer for Science Education and The Journal of Research in Science Teaching. He worked on the Delaware Teacher Standards, Delaware Science Assessment, National Board of Teacher Certification, and AAAS Project 2061 Curriculum Evaluation programs.

Senior Writer

John Southard received his undergraduate degree from the Massachusetts Institute of Technology in 1960 and his doctorate in geology from Harvard University in 1966. After a National Science Foundation postdoctoral fellowship at the California Institute of Technology, he joined the faculty at the Massachusetts Institute of Technology, where he is currently Professor of Geology Emeritus. He was awarded the MIT School of Science teaching prize in 1989 and was one of the first cohorts of first MacVicar Fellows at MIT, in recognition of excellence in undergraduate teaching. He has taught numerous undergraduate courses in introductory geology, sedimentary geology, field geology, and environmental Earth Science both at MIT and in Harvard's adult education program. He was editor of the Journal of Sedimentary Petrology from 1992 to 1996, and he continues to do technical editing of scientific books and papers for SEPM, a professional society for sedimentary geology. Dr. Southard received the 2001 Neil Miner Award from the National Association of Geoscience Teachers.

Project Director/Curriculum Designer

Colin Mably has been a key curriculum developer for several NSF-supported national curriculum projects. As learning materials designer to the American Geological Institute, he has directed the design and development of the IES curriculum modules and also training workshops for pilot and field-test teachers.

Teacher's Edition

Project Team
Marcus Milling
Executive Director - AGI, VA
Michael Smith
Principal Investigator
Director of Education - AGI, VA
Colin Mably
Project Director/Curriculum Designer
Educational Visions, MD
Fred Finley
Project Evaluator
University of Minnesota, MN
Lynn Lindow
Pilot Test Evaluator
University of Minnesota, MN
Harvey Rosenbaum
Field Test Evaluator
Montgomery School District, MD
Ann Benbow
Project Advisor - American Chemical Society, DC
Robert Ridky
Original Project Director
University of Maryland, MD
Chip Groat
Original Principal Investigator
University of Texas - El Paso, TX
Marilyn Suiter
Original Co-principal Investigator
AGI, VA
William Houston
Project Manager
Eric Shih - Project Assistant

Original and Contributing Authors
Oceans
George Dawson
Florida State University, FL
Joseph F. Donoghue
Florida State University, FL
Ann Benbow
American Chemical Society
Michael Smith
American Geological Institute
Soil
Robert Ridky
University of Maryland, MD
Colin Mably - LaPlata, MD
John Southard
Massachusetts Institute of Technology, MA
Michael Smith
American Geological Institute
Fossils
Robert Gastaldo
Colby College, ME
Colin Mably - LaPlata, MD
Michael Smith
American Geological Institute
Climate and Weather
Mike Mogil
How the Weather Works, MD

Ann Benbow
American Chemical Society
Joe Moran
American Meteorological Society
Michael Smith
American Geological Institute
Energy Resources
Laurie Martin-Vermilyea
American Geological Institute
Michael Smith
American Geological Institute
Dynamic Planet
Michael Smith
American Geological Institute
Rocks and Landforms
Michael Smith
American Geological Institute
Water as a Resource
Ann Benbow
American Chemical Society
Michael Smith
American Geological Institute
Materials and Minerals
Mary Poulton
University of Arizona, AZ
Colin Mably - LaPlata, MD
Michael Smith
American Geological Institute

Content Reviewers
Louis Bartek
University of North Carolina
Gary Beck - BP Exploration
Steve Bergman
University of Texas-Dallas
Joseph Bishop
Johns Hopkins University/NOAA
Kathleen Carrado
Argonne National Laboratory
Sandip Chattopadhyay
R.S. Kerr Environmental Research Center
Bob Christman
Western Washington University
Donald Conte
California University of California
Norbert E. Cygan - AAPG
Tom Dignes
Mobil Technology Corporation
Neil M. Dubrovsky
United States Geological Survey
Robert J. Finley
Illinois State Geological Survey
Anke Friedrich
California Institute of Technology
Rick Fritz - AAPG
Frank Hall - University of New Orleans
David Hawkins - Denison University
Martha House
California Institute of Technology
Travis Hudson
American Geological Institute
Allan P. Juhas - SEG

Dennis Lamb - Penn State
Donald Lewis - Happy Valley, CA
Kate Madin
Woods Hole Oceanographic Institute
John Madsen - University of Delaware
Carol Mankiewicz - Beloit College
Clyde J. Northrup
Boise State University
Lois K. Ongley, PhD - Bates College
Bruce Pivetz
ManTech Environmental Research Services
Eleanora I. Robbins
United States Geological Survey
Rob Ross
Paleontological Research Institution
Audrey Rule - Boise State University
Lou Solebello - Macon, GA
Steve Stanley - Johns Hopkins University
Sarah Tebbens - University of South Florida
Bob Tilling
United States Geological Survey
Michael Velbel
Michigan State University
Don Woodrow
Hobart and William Smith Colleges

Pilot Test Teachers
Debbie Bambino - Philadelphia, PA
Barbara Barden - Rittman, OH
Louisa Bliss - Bethlehem, NH
Mike Bradshaw - Houston TX
Greta Branch - Reno, NV
Garnetta Chain - Piscataway, NJ
Roy Chambers - Portland, OR
Laurie Corbett - Sayre, PA
James Cole - New York, NY
Collette Craig - Reno, NV
Anne Douglas - Houston, TX
Jacqueline Dubin - Roslyn, PA
Jane Evans - Media, PA
Gail Gant - Houston, TX
Joan Gentry - Houston, TX
Pat Gram - Aurora, OH
Robert Haffner - Akron, OH
Joe Hampel - Swarthmore, PA
Wayne Hayes - West Green, GA
Mark Johnson - Reno, NV
Cheryl Joloza - Philadelphia, PA
Jeff Luckey - Houston, TX
Karen Luniewski
Reistertown, MD
Cassie Major - Plainfield, VT
Carol Miller - Houston, TX
Melissa Murray - Reno, NV
Mary-Lou Northrop

Tracey Oliver - Philadelphia, PA
Nicole Pfister - Londonderry, VT
Beth Price - Reno, NV
Joyce Ramig - Houston, TX
Julie Revilla - Woodbridge, VA
Steve Roberts - Meredith, NH
Cheryl Skipworth - Philadelphia, PA
Brent Stenson - Valdosta, GA
Elva Stout - Evans, GA
Regina Toscani - Philadelphia, PA
Bill Waterhouse - North Woodstock, NH
Leonard White - Philadelphia, PA
Paul Williams - Lowerford, VT
Bob Zafran - San Jose, CA
Missi Zender - Twinsburg, OH

Field Test Teachers
Eric Anderson - Carson City, NV
Katie Bauer - Rockport, ME
Kathleen Berdel - Philadelphia, PA
Wanda Blake - Macon, GA
Beverly Bowers - Mannington, WV
Rick Chiera - Monroe Falls, OH
Don Cole - Akron, OH
Patte Cotner - Bossier City, LA
Johnny DeFreese - Haughton, LA
Mary Devine - Astoria, NY
Cheryl Dodes - Queens, NY
Brenda Engstrom - Warwick, RI
Lisa Gioe-Cordi - Brooklyn, NY
Pat Gram - Aurora, OH
Mark Johnson - Reno, NV
Chicory Koren - Kent, OH
Marilyn Krupnick - Philadelphia, PA

Melissa Loftin - Bossier City, LA
Janet Lundy - Reno, NV
Vaughn Martin - Easton, ME
Anita Mathis - Fort Valley, GA
Laurie Newton - Truckee, NV
Debbie O'Gorman - Reno, NV
Joe Parlier - Barnesville, GA
Sunny Posey - Bossier City, LA
Beth Price - Reno, NV
Stan Robinson - Mannington, WV
Mandy Thorne - Mannington, WV
Marti Tomko - Westminster, MD
Jim Trogden - Rittman, OH
Torri Weed - Stonington, ME
Gene Winegart - Shreveport, LA
Dawn Wise - Peru, ME
Paula Wright - Gray, GA

IMPORTANT NOTICE

The *Investigating Earth Systems*™ series of modules is intended for use by students under the direct supervision of a qualified teacher. The experiments described in this book involve substances that may be harmful if they are misused or if the procedures described are not followed. Read cautions carefully and follow all directions. Do not use or combine any substances or materials not specifically called for in carrying out experiments. Other substances are mentioned for educational purposes only and should not be used by students unless the instructions specifically indicate.

The materials, safety information, and procedures contained in this book are believed to be reliable. This information and these procedures should serve only as a starting point for classroom or laboratory practices, and they do not purport to specify minimal legal standards or to represent the policy of the American Geological Institute. No warranty, guarantee, or representation is made by the American Geological Institute as to the accuracy or specificity of the information contained herein, and the American Geological Institute assumes no responsibility in connection therewith. The added safety information is intended to provide basic guidelines for safe practices. It cannot be assumed that all necessary warnings and precautionary measures are contained in the printed material and that other additional information and measures may not be required.

This work is based upon work supported by the National Science Foundation under Grant No. 9353035 with additional support from the Chevron Corporation. Any opinions, findings, and conclusions or recommendations expressed in this publication are those of the authors and do not necessarily reflect the views of the National Science Foundation or the Chevron Corporation. Any mention of trade names does not imply endorsement from the National Science Foundation or the Chevron Corporation.

Teacher's Edition

The American Geological Institute and Investigating Earth Systems

Imagine more than 500,000 Earth scientists worldwide sharing a common voice, and you've just imagined the mission of the American Geological Institute. Our mission is to raise public awareness of the Earth sciences and the role that they play in mankind's use of natural resources, mitigation of natural hazards, and stewardship of the environment. For more than 50 years, AGI has served the scientists and teachers of its Member Societies and hundreds of associated colleges, universities, and corporations by producing Earth science educational materials, *Geotimes*–a geoscience news magazine, GeoRef–a reference database, and government affairs and public awareness programs.

So many important decisions made every day that affect our lives depend upon an understanding of how our Earth works. That's why AGI created *Investigating Earth Systems*. In your *Investigating Earth Systems* classroom, you'll discover the wonder and importance of Earth science. As you investigate minerals, soil, or oceans — do field work in nearby beaches, parks, or streams, explore how fossils form, understand where your energy resources come from, or find out how to forecast weather — you'll gain a better understanding of Earth science and its importance in your life.

We would like to thank the National Science Foundation and the AGI Foundation Members that have been supportive in bringing Earth science to students. The Chevron Corporation provided the initial leadership grant, with additional contributions from the following AGI Foundation Members: Anadarko Petroleum Corp., The Anschutz Foundation, Baker Hughes Foundation, Barrett Resources Corp., BPAmoco Foundation, Burlington Resources Foundation, CGG Americas, Inc., Conoco Inc., Consolidated Natural Gas Foundation, Diamond Offshore Co., EEX Corp., ExxonMobil Foundation, Global Marine Drilling Co., Halliburton Foundation, Inc., Kerr McGee Foundation, Maxus Energy Corp., Noble Drilling Corp., Occidental Petroleum Charitable Foundation, Parker Drilling Co., Phillips Petroleum Co., Santa Fe Snyder Corp., Schlumberger Foundation, Shell Oil Company Foundation, Southwestern Energy Co., Texaco, Inc., Texas Crude Energy, Inc., Unocal Corp. USX Foundation (Marathon Oil Co.).

We at AGI wish you success in your exploration of the Earth System!

Michael J. Smith
Director of Education, AGI

Marcus E. Milling
Executive Director, AGI

Developing Investigating Earth Systems

Welcome to *Investigating Earth Systems (IES)!* IES was developed through funding from the National Science Foundation and the American Geological Institute Foundation. Classroom teachers, scientists, and thousands of students across America helped to create *IES*. In the 1997-98 school year, scientists and curriculum developers drafted nine *IES* modules. They were pilot tested by 43 teachers in 14 states from Washington to Georgia. Faculty from the University of Minnesota conducted an independent evaluation of the pilot test in 1998, which was used to revise the program for a nationwide field test during the 1999-2000 school year. A comprehensive evaluation of student learning by a professional field-test evaluator showed that *IES* modules led to significant gains in student understanding of fundamental Earth science concepts. Field-test feedback from 34 teachers and content reviews from 33 professional Earth scientists were used to produce the commercial edition you have selected for your classroom.

Inquiry and the interrelation of Earth's systems form the backbone of *IES*. Often taught as a linear sequence of events called "the scientific method," inquiry underlies all scientific processes and can take many different forms. It is very important that students develop an understanding of inquiry processes as they use them. Your students naturally use inquiry processes when they solve problems. Like scientists, students usually form a question to investigate after first looking at what is observable or known. They predict the most likely answer to a question. They base this prediction on what they already know to be true. Unlike professional scientists, your students may not devote much thought to these processes. In order to be objective, students must formally recognize these processes as they do them. To make sure that the way they test ideas is fair, scientists think very carefully about the design of their investigations. This is a skill your students will practice throughout each *IES* module.

All *Investigating Earth Systems* modules also encourage students to think about the Earth as a system. Upon completing each investigation they are asked to relate what they have learned to the Earth Systems (see the *Earth System Connection* sheet in the **Appendix**). Integrating the processes of the biosphere, geosphere, hydrosphere, and atmosphere will open up a new way of looking at the world for most students. Understanding that the Earth is dynamic and that it affects living things, often in unexpected ways, will engage them and make the topics more relevant.

We trust that you will find the Teacher's Edition that accompanies each student module to be useful. It provides **Background Information** on the concepts explored in the module, as well as strategies for incorporating inquiry and a systems-based approach into your classroom. Enjoy your investigation!

Teacher's Edition

Investigating Earth Systems Modules

Climate and Weather

Dynamic Planet

Energy Resources

Fossils

Materials and Minerals

Oceans

Rocks and Landforms

Soil

Water as a Resource

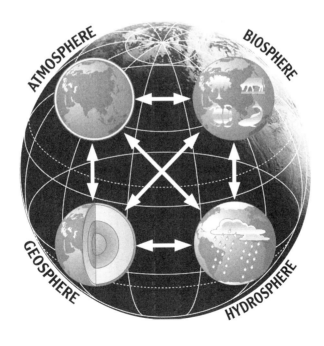

Investigating Earth Systems: Correlation to the National Science Education Standards

National Science Education Content Standards Grades 5 – 8	Soil	Rocks and Landforms	Oceans	Climate and Weather	Dynamic Planet	Materials and Minerals	Energy Resources	Water as a Resource	Fossils
UNIFYING CONCEPTS AND PROCESSES									
System, order, and organization	•	•	•	•	•	•	•	•	•
Evidence, models, and explanation	•	•	•	•	•	•	•	•	•
Constancy, change, and measurement	•	•	•	•	•	•	•	•	•
Evolution and equilibrium		•	•	•	•			•	•
Form and function									•
SCIENCE AS INQUIRY									
Identify questions that can be answered through scientific investigations	•	•	•	•	•	•	•	•	•
Design and conduct scientific investigations	•	•	•	•	•	•	•	•	•
Use tools and techniques to gather, analyze, and interpret data	•	•	•	•	•	•	•	•	•
Develop descriptions, explanations, predictions and models based on evidence	•	•	•	•	•	•	•	•	•
Think critically and logically to make the relationships between evidence and explanation	•	•	•	•	•	•	•	•	•
Recognize and analyze alternative explanations and predictions	•	•	•	•	•	•	•	•	•
Communicate scientific procedures and explanations	•	•	•	•	•	•	•	•	•
Use mathematics in all aspects of scientific inquiry	•	•	•	•	•	•	•	•	•
Understand scientific inquiry	•	•	•	•	•	•	•	•	•
PHYSICAL SCIENCE									
Properties and Changes of Properties in Matter	•	•	•			•	•	•	
Motions and Forces	•		•						
Transfer of Energy		•	•	•	•	•	•	•	
LIFE SCIENCE									
Populations and Ecosystems			•				•	•	•
Diversity and Adaptation of Organisms			•		•				•

Teacher's Edition

Investigating Earth Systems: Correlation to the National Science Education Standards

National Science Education Content Standards Grades 5 – 8

	Soil	Rocks and Landforms	Oceans	Climate and Weather	Dynamic Planet	Materials and Minerals	Energy Resources	Water as a Resource	Fossils
EARTH AND SPACE SCIENCE									
Structure of the Earth system	•	•	•	•	•	•	•	•	•
Earth's History	•	•	•	•	•	•	•	•	•
Earth in the Solar System			•	•	•		•	•	
SCIENCE AND TECHNOLOGY									
Abilities of technological design	•	•	•	•	•	•	•	•	•
Understandings about science and technology		•	•			•	•	•	
SCIENCE IN PERSONAL AND SOCIAL PERSPECTIVES									
Personal health	•							•	
Populations, resources, and environment	•					•	•	•	
Natural Hazards		•		•	•	•			
Risks and benefits					•		•		
Science and technology in society	•	•	•	•	•	•	•	•	•
HISTORY AND NATURE OF SCIENCE									
Science as a human endeavor	•	•	•	•	•	•	•	•	•
Nature of science	•	•	•	•	•	•	•	•	•
History of science			•		•				•

Investigating Earth Systems: Climate and Weather

Using Investigating Earth Systems Features in Your Classroom

1. Pre-assessment

Designed under the umbrella framework of "science for all students," meaning that all students should be able to engage in inquiry and learn core science concepts, *Investigating Earth Systems* helps you to tailor instruction to meet your students' needs. A crucial first step in this framework is to ascertain what knowledge, experience, and understanding your students bring to their study of a module. The pre-assessment consists of four questions geared to the major concepts and understandings targeted in the unit. Students write and draw what they know about the major topics and concepts. This information is recorded and shared in an informal discussion prior to engaging in hands-on inquiry. The discussion enables students to recognize how much there is to learn and appreciate, and that by exploring the unit together, the entire classroom can emerge from the experience with a better understanding of core concepts and themes. Students' responses provide crucial pre-assessment data for you. By examining their written work and probing for further detail during the classroom conversation, you can identify strengths and weaknesses in students' understandings, as well as their abilities to communicate that understanding to others. It is important that the pre-assessment not be viewed as a test, and that judgments about the accuracy of responses be evaluated in writing or through your comments during the conversation. The goal is to ascertain and probe, not judge, and to create a safe classroom environment in which students feel comfortable sharing their ideas and knowledge. Students revisit these pre-assessment questions informally throughout the unit. At the end of the unit, students respond to the same four questions in the section called **Back to the Beginning**. The pre-assessment thus helps you and your students to make judgments about their growth in understanding and ability throughout the module.

2. The Earth System

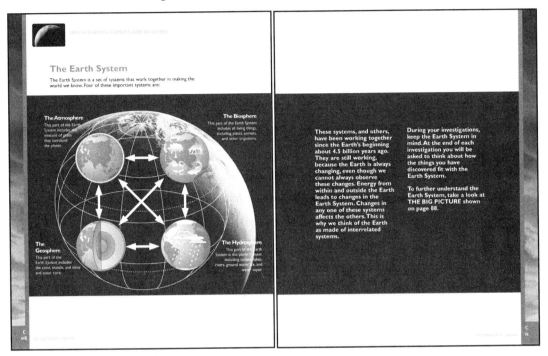

National Science Education Standards link...

"A major goal of science in the middle grades is for students to develop an understanding of Earth (and the solar system) as a set of closely coupled systems. The idea of systems provides a framework in which students can investigate the four major interacting components of the Earth System – geosphere (crust, mantle, and core), hydrosphere (water), atmosphere (air), and the biosphere (the realm of living things)."

NSES content standard D "Developing Student Understanding" (pages 158-159)

Understanding the Earth system is an overall goal of the *Investigating Earth Systems* series. It is a difficult and complex set of concepts to grasp, because it is inferred rather than observed directly. Yet even the smallest component of Earth science can be linked to the Earth system. As your students progress through each module, an increasing number of connections with the Earth system will arise. Your students may not, however, immediately see these connections. At the end of every investigation, they will be asked to link what they have discovered with ideas about the Earth system. They will also be asked to write about this in their journals. A **Blackline Master** (*Earth System Connection* sheet) is available in each Teacher's Edition. Students can use this to record connections that they make as they complete each investigation. At the very end of the module they will be asked to review everything they have learned in relation to the Earth system. The aim is for students to have a working understanding of the Earth System by the time they complete grade 8. They will need your help accomplishing this.

For example, in *Investigating Rocks and Landforms*, students work with models to simulate Earth processes, such as erosion of stream sediment and deposition of that sediment on floodplains and in deltas. Changes in inputs in one part of the system (say rainfall, from the atmosphere), affect other parts of the system (stream flows, erosion on river bends, amount of sediment carried by the stream, and deposition of sediment on floodplains or in deltas). These changes affect, in turn, other parts of the system (for example, floods that affect human populations, i.e., the biosphere). In the same module, students explore the rock record within their community and develop understandings about how interactions between the hydrosphere, atmosphere, geosphere, and biosphere change the landscape over time. These are just some of the many ways that *Investigating Earth Systems* modules foster and promote student thinking about the dynamic nature and interactions of Earth systems—biosphere, geosphere, atmosphere, and hydrosphere.

3. Introducing Inquiry Processes

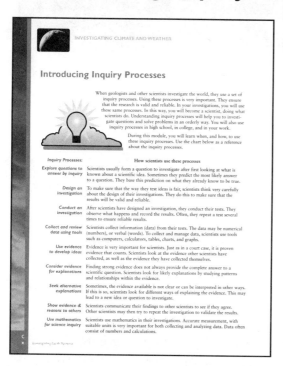

Inquiry is at the heart of *Investigating Earth Systems*. That is why each module title begins with the title "Investigating." In the National Science Education Standards, inquiry is the first Content Standard. NSES then lists a range of points about inquiry. These fundamental components of inquiry were written into the list shown at the beginning of each student module. It is very important that students be reminded of the steps in the inquiry process as they perform them. Inquiry depends on active student participation. Ideas on how to make inquiry successful in the classroom appear throughout the modules and in the "Managing Inquiry in Your *Investigating Earth Systems* Classroom" section of this Teacher's Edition.

It is very important that students develop an understanding of the inquiry processes as they use them. Stress the importance of inquiry processes as they occur in your investigations. Provoke students to think about why these processes are important. Collecting good data, using evidence, considering alternative explanations, showing evidence to others, and using mathematics are all essential to *IES*. Use examples to demonstrate these processes whenever possible. At the end of every investigation, students are asked to reflect on the scientific inquiry processes they used. Refer students to the list of inquiry processes on page Cx of the Student Book as they think about scientific inquiry and answer the questions.

4. Introducing the Module

Each *IES* module begins with photographs and questions. This is an introduction to the module for your students. It is designed to give them a brief overview of the content of the module and set their investigations into a relevant and meaningful context. Students will have had a variety of experiences with the content of the module. This is an opportunity for them to offer some of their own experiences in a general discussion, using these questions as prompts. This section of each *IES* module follows the pre-assessment, where students spend time thinking about what they already know about the content of the module. The photographs and questions can be used to focus the students' thinking.

The ideas students share in the introduction to the module provide you with additional pre-assessment data. The experiences they describe and the way in which they are discussed will alert you to their general level of understanding about these topics. To encourage sharing and to provide a record, teachers find it useful to quickly summarize the main points that emerge from discussion. You can do this on the chalkboard or flipchart for all to see. This can be displayed as students work through the module and added to with each new experience. For your own assessment purposes, it will be useful to keep a record of these early indicators of student understanding.

5. Key Question

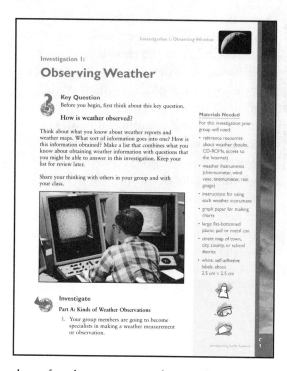

Each *Investigating Earth Systems* investigation begins with a **Key Question** – an open-ended question that gives teachers the opportunity to explore what their students know about the central concepts of the activity. Uncovering students' thinking (their prior knowledge) and exposing the diversity of ideas in the classroom are the first steps in the learning cycle. One of the most fundamental principles derived from many years of research on student learning is that:

"Students come to the classroom with preconceptions about how the world works. If their initial understanding is not engaged, they may fail to grasp the new concepts and information that are taught, or they may learn them for the purposes of a test but revert to their preconceptions outside the classroom." (*How People Learn: Bridging Research and Practice*, National Research Council, 1999, P. 10.)

This principle has been illustrated through the *Private Universe* series of videotapes that show Harvard graduates responding to basic science questions in much the same way that fourth grade students do. Although the videotapes revealed that the Harvard graduates used a more sophisticated vocabulary, the majority held onto the same naïve, incorrect conceptions of elementary school students. Research on learning suggests that the belief systems of students who are not confronted with what they believe and adequately shown why they should give up that belief system remain intact. Real learning requires confronting one's beliefs and testing them in light of competing explanations.

Drawing out and working with students' preconceptions is important for learners. In *Investigating Earth Systems*, the **Key Question** is used to ascertain students' prior knowledge about the key concept or Earth science processes or events explored in the activity. Students verbalize what they think about the age of the Earth, the causes of volcanoes, or the way that the landscape changes over time before they embark on an activity designed to challenge and test these beliefs. A brief discussion about the diversity of beliefs in the classroom makes students consider how their ideas compare to others and the evidence that supports their view of volcanoes, earthquakes, or seasons.

Teacher's Edition

The **Key Question** is not a conclusion, but a lead into inquiry. It is not designed to instantly yield the "correct answer" or a debate about the features of the question, or to bring closure. The activity that follows will provide that discussion as students analyze and discuss the results of inquiry. Students are encouraged to record their ideas in words and/or drawings to ensure that they have considered their prior knowledge. After students discuss their ideas in pairs or in small groups, teachers activate a class discussion. A discussion with fellow students prior to class discussion may encourage students to exchange ideas without the fear of personally giving a "wrong answer." Teachers sometimes have students exchange papers and volunteer responses that they find interesting.

Some teachers prefer to have students record their responses to these questions. They then call for volunteers to offer ideas up for discussion. Other teachers prefer to start with discussion by asking students to volunteer their ideas. In either situation, it is important that teachers encourage the sharing of ideas by not judging responses as "right" or "wrong." It is also important that teachers keep a record of the variety of ideas, which can be displayed in the classroom (on a sheet of easel pad paper or on an overhead transparency) and referred to as students explore the concepts in the module. Teachers often find that they can group responses into a few categories and record the number of students who hold each idea. The photograph in each **Key Question** section was designed to stimulate student thinking and help students to make the specific kinds of connections emphasized in each activity.

6. Investigate

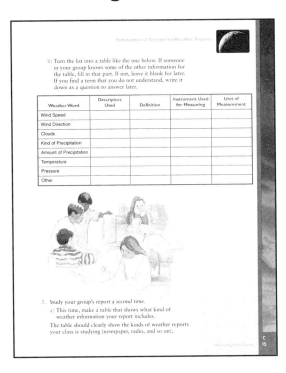

Investigating Earth Systems is a hands-on, minds-on curriculum. In designing *Investigating Earth Systems*, we were guided by the research on learning, which points out how important ***doing*** Earth Science is to ***learning*** Earth Science. Testing of *Investigating Earth Systems* activities by teachers across America provided critical testimonial and quantitative measures of the importance of the activities to student learning. In small groups and as a class, students take part in doing hands-on experiments, participating in field work, or searching for answers using the Internet and reference materials. **Blackline Masters** are included in the Teacher's Editions for any maps or illustrations that are essential for students to complete the activity.

Investigating Earth Systems: Climate and Weather

Each part of an *Investigating Earth Systems* investigation, as well as the sequence of activities within a module, moves from concrete to abstract. Hands-on activities provide the basis for exploring student beliefs about how the world works and to manipulate variables that affect the outcomes of experiments, models, or simulations. Later in each activity, formal labels are applied to concepts by introducing terminology used to describe the processes that students have explored through hands-on activity. This flow from concrete (hands-on) to abstract (formal explanations) is progressive – students begin to develop their own explanations for phenomena by responding to questions within the **Investigate** section.

Each activity has instructions for each part of the investigation. Materials kits are available for purchase, but you will also need to obtain some resources from outside suppliers, such as topographic and geologic maps of your community, state, or region. The *Investigating Earth Systems* web site will direct you to sources where you can gather such materials.

Most **Investigate** activities will require between one and two class periods. The variety of school schedules and student needs makes it difficult to predict exactly how much time your class will need. For example, if students need to construct a graph for part of an investigation, and the students have never been exposed to graphing, then this investigation may require additional time and could become part of a mathematics lesson.

The most challenging aspect of *Investigating Earth Systems* for teachers to "master" is that the **Investigate** section of each activity has been designed to be student-driven. Students learn more when they have to struggle to "figure things out" and work in collaborative groups to solve problems as a team. Teachers will have to resist the temptation to provide the answers to students when they get "stuck" or hung up on part of a problem. Eventually, students learn that while they can call upon their teacher for assistance, the teacher is not going to "show them the answer." Field testing of *Investigating Earth Systems* revealed that teachers who were most successful in getting their students to solve problems as a team were patient with this process and steadfast in their determination to act as facilitators of learning during the **Investigate** portion of activities. As one teacher noted, "My response to questions during the investigation was like a mantra, 'What do you think you need to do to solve this?' My students eventually realized that although I was there to provide guidance, they weren't going to get the solution out of me."

Another concern that many teachers have when examining *Investigating Earth Systems* for the first time is that their students do not have the background knowledge to do the investigations. They want to deliver a lecture about the phenomena before allowing students to do the investigation. Such an approach is common to many traditional programs and is inconsistent with the pedagogical theory used to design *Investigating Earth Systems*. The appropriate place for delivering a lecture or reading text in *Investigating Earth Systems* is following the investigation, not preceding it.

Teacher's Edition

For example, suppose a group of students has been asked to interpret a map. The traditional approach to science education is for the teacher to give a lecture or assign a reading, "How to Interpret Maps," then give students practice reading maps. *Investigating Earth Systems* teachers recognize that while students may lack some specific skills (reading latitude and longitude, for example), within a group of four students, it is not uncommon for at least one of the students to have a vital skill or piece of knowledge that is required to solve a problem. The one or two students who have been exposed to (or better yet, understood) latitude and longitude have the opportunity to shine within the group by contributing that vital piece of information or demonstrating a skill. That's how scientific research teams work – specialists bring expertise to the group, and by working together, the group achieves something that no one could achieve working alone. The **Investigate** section of *Investigating Earth Systems* is modeled in the spirit of the scientific research team.

7. Inquiry

Inquiry is the first content standard in the National Science Education Standards (NSES). The American Association for the Advancement of Science's (AAAS) Benchmarks for Science Literacy also places considerable emphasis on scientific inquiry (see excerpts on the following page). *IES* has been designed to remind students to reflect on inquiry processes as they carry out their investigations. The student journal is an important tool in helping students to develop these understandings. In using the journal, students are modeling what scientists do. Your students are young scientists as they investigate Earth science questions. Encourage your students to think of themselves in this way and to see their journals as records of their investigations.

Inquiry
Representing Information

Communicating findings to other scientists is very important in scientific inquiry. In this investigation it is important for you to find good ways of showing what you learned to others in your class. Be sure your maps and displays are clearly labeled and well organized.

An icon was developed to draw students' attention to brief descriptions of inquiry processes in the margins of the student module. The icon and explanations provide opportunities to direct students' attention to what they are doing, and thus serve as an important metacognitive tool to stimulate thinking about thinking.

National Science Education Standards link...

Content Standard A
As a result of activities in grades 5-8, all students should develop:
- Abilities necessary to do scientific inquiry
- Understandings about scientific inquiry

Abilities Necessary to do Scientific Inquiry
- Identify questions that can be answered through scientific investigations
- Use appropriate tools and techniques to gather, analyze, and interpret data
- Develop descriptions, explanations, predictions, and models using evidence
- Think critically and logically to make the relationships between evidence and explanations
- Recognize and analyze alternative explanations and predictions
- Communicate scientific procedures and explanations
- Use mathematics in all aspects of scientific inquiry

(From National Science Education Standards, pages 145-148)

Benchmarks for Science Literacy link...

The Nature of Science Inquiry: Grades 6 through 8
- At this level, students need to become more systematic and sophisticated in conducting their investigations, some of which may last for several weeks. That means closing in on an understanding of what constitutes a good experiment. The concept of controlling variables is straightforward, but achieving it in practice is difficult. Students can make some headway, however, by participating in enough experimental investigations (not to the exclusion, of course, of other kinds of investigations) and explicitly discussing how explanation relates to experimental design.

- Student investigations ought to constitute a significant part—but only a part—of the total science experience. Systematic learning of science concepts must also have a place in the curriculum, for it is not possible for students to discover all the concepts they need to learn, or to observe all of the phenomena they need to encounter, solely through their own laboratory investigations. And even though the main purpose of student investigations is to help students learn how science works, it is important to back up such experience with selected readings. This level is a good time to introduce stories (true and fictional) of scientists making discoveries – not just world-famous scientists, but scientists of very different backgrounds, ages, cultures, places, and times.

(From Benchmarks for Science Literacy, page 12)

Teacher's Edition

8. Digging Deeper

This section provides text, illustrations, data tables, and photographs that give students greater insight into the concepts explored in the activity. Teachers often assign **As You Read** questions as homework to guide students to think about the major ideas in the text. Teachers can also select questions to use as quizzes, rephrasing the questions into multiple choice or "true/false" formats. This provides assessment information about student understanding and serves as a motivational tool to ensure that students complete the reading assignment and comprehend the main ideas.

This is the stage of the activity that is most appropriate for teachers to explain concepts to students in whole-class lectures or discussions. References to **Blackline Masters** are available throughout the Teacher's Edition. They refer to illustrations from the textbook that teachers may photocopy and distribute to students or make overhead transparencies for lectures or presentations.

9. Review and Reflect

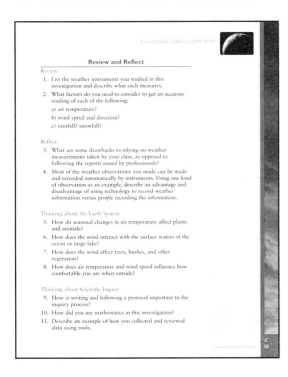

Questions in this feature ask students to use the key principles and concepts introduced in the activity. Students are sometimes presented with new situations in which they are asked to apply what they have learned. The questions in this section typically require higher-order thinking and reasoning skills than the **As You Read** questions. Teachers can assign these questions as homework, or have students complete them in groups during class. Assigning them as homework economizes time available in class, but has the drawback of making it difficult for students to collectively revisit the understanding that they developed as they worked through the concepts as a group during the investigation. A third alternative is, of course, to assign the work individually in class. When students work through application problems in class, teachers have the opportunity to interact with students at a critical juncture in their learning – when they may be just on the verge of "getting it."

Review and Reflect prompts students to think about what they have learned, how their work connects with the Earth system, and what they know about scientific inquiry. Another one of the important principles of learning used to guide the selection of content in *Investigating Earth Systems* was that:

"To develop competence in an area of inquiry, students must (a) have a deep foundation of factual knowledge, (b) understand facts and ideas in the context of a conceptual framework, and (c) organize knowledge in ways that facilitate retrieval and application." (*How People Learn: Bridging Research and Practice* National Research Council, 1999, P. 12.)

Reflecting on one's learning and one's thinking is an important metacognitive tool that makes students examine what they have learned in the activity and then think critically about the usefulness of the results of their inquiry. It requires students to take stock of their learning and evaluate whether or not they really understand "how it fits into the Big Picture." It is important for teachers to guide students through this process with questions such as "What part of your work demonstrates that you know and can do scientific inquiry? How does what you learned help you to better understand the Earth system? How does your work contribute or relate to the concepts of the Big Picture at the end of the module?"

10. Final Investigation: Putting It All Together

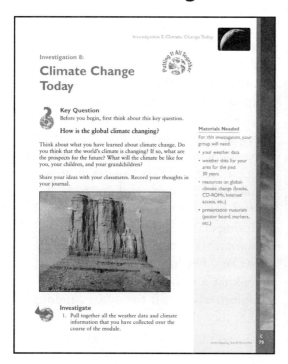

In the final investigation in each *Investigating Earth Systems* module, your students will apply all the knowledge they have about the topics explored to solve a practical problem or situation. Requiring students to apply all they have gained toward a specific outcome should serve as the main assessment information for the module. A sample assessment rubric is provided in the back of this Teacher's Edition. Whatever rubric you employ, it is important that you share this with students at the outset of the final investigation so that they understand the criteria upon which their work will be judged.

The instructions provided to students are purposely open-ended, but can be completed to various levels depending upon how much knowledge students apply. During the final investigation, your role is to be a participant observer, moving from group to group, noticing how students go about the investigation and how they are applying the experience and understanding they have gained from the module.

11. Reflecting

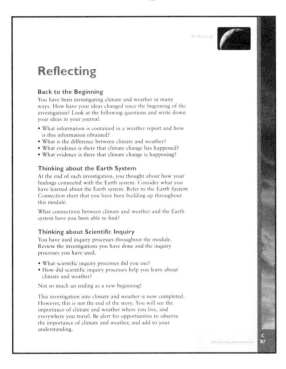

Now that students are at the end of the module, they are provided with questions that ask them to reflect upon all that they have learned about Earth science, inquiry, and the Earth system. The first set of questions (**Back to the Beginning**) are the same questions used in the pre-assessment. Teachers often ask students to revisit their initial responses and provide new answers to demonstrate how much they have learned.

12. The Big Picture

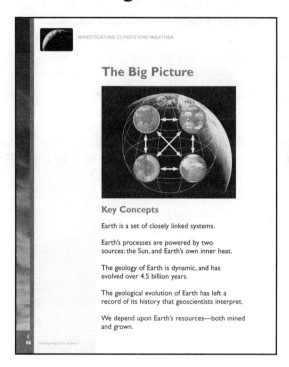

The five key concepts below underlie Earth science in general and *Investigating Earth Systems* in particular. Collectively, the nine modules in the *Investigating Earth Systems* series are designed to help students understand each of these concepts by the time they complete grade 8. Many of the concepts that underlie the Big Picture may be difficult for students to grasp easily. As students develop their ideas through inquiry-based investigations, you can help them to make connections with these key scientific concepts. As a reminder of the importance of the major understandings, the Student Book has a copy of the Big Picture in the back of the book near the **Glossary**.

Be on the lookout for chances to remind students that:
- Earth is a set of closely linked systems.
- Earth's processes are powered by two sources: the Sun and Earth's own inner heat.
- The geology of Earth is dynamic, and has evolved over 4.5 billion years.
- The geological evolution of Earth has left a record of its history that geoscientists interpret.
- We depend upon Earth's resources—both mined and grown.

13. Glossary

Words that may be new or unfamiliar to students are defined and explained in the **Glossary** of the Student Book. Teachers use their own judgment about selecting the terms that appear in the **Glossary** that are most important for their students to learn. Teachers typically use discretion and consider their state and local guidelines for science content understanding when assigning importance to particular vocabulary, which in most cases is very likely to be a small subset of all the scientific terms introduced in each module and defined in the **Glossary**.

References

How People Learn: Bridging Research and Practice (1999) Suzanne Donovan, John Bransford, and James Pellegrino, editors. National Academy Press, Washington, DC. 78 pages. The report is also available online at www.nap.edu.

Using the Investigating Earth Systems Web Site

www.agiweb.org/ies

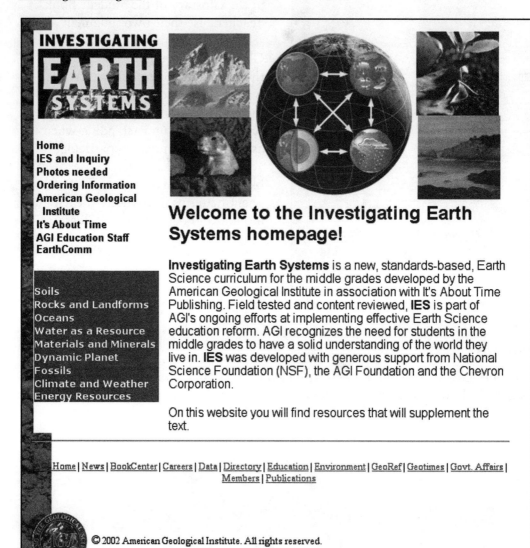

The *Investigating Earth Systems* web site has been designed for teachers and students.
- Each *Investigating Earth Systems* module has its own web page that has been designed specifically for the content addressed within that module.
- Module web sites are broken down by investigation and also contain a section with links to relevant resources that are useful for the module.
- Each investigation is divided into materials and supplies needed, **Background Information,** and links to resources that will help you and your students to complete the investigation.

Teacher's Edition

Enhancing Teacher Content Knowledge

Each *Investigating Earth Systems* module has a specific web page that will help teachers to gather further **Background Information** about the major topics covered in each activity.

Example from *Investigating Rocks and Landforms* – Investigation 1
Different Types of Rock

To learn more about different types of rocks, visit the following web sites:

- What are the basic types of rock?, Rogue Community College
This site lists the basic descriptions of sedimentary, metamorphic and igneous rocks. Detailed information on each type of rock is also available.
(http://www.jersey.uoregon.edu/~mstrick/AskGeoMan/geoQuerry13.html)

1. Sedimentary Rocks:
- Sedimentary Rocks, University of Houston
Detailed description of the composition, classification, and formation of sedimentary rocks.
(http://ucaswww.mcm.uc.edu/geology/maynard/INTERNETGUIDE/appendf.htm)
- Image Gallery for Geology, University of North Carolina
See more examples of sedimentary rocks.
(http://www.geosci.unc.edu/faculty/glazner/Images/SedRocks/SedRocks.html)
- Sedimentary Rocks Laboratory, Georgia Perimeter College
Read a thorough discussion of clastic, chemical, and organic sedimentary rocks. Illustrations accompany each description.
(http://www.gpc.peachnet.edu/~pgore/geology/historical_lab/sedrockslab.php)
- Textures and Structures of Sedimentary Rocks, Duke University
View a collection of slides of different sedimentary rocks as either outcrops or thin sections viewed through a microscope.
(http://www.geo.duke.edu/geo41/seds.htm)
- Sedimentary Rocks, Washington State University
Learn more about sedimentary processes, environments of deposition in relation to different sedimentary rocks. Topics covered include depositional environments, chemical or mechanical weathering, deposition and lithification, and classification.
(http://www.wsu.edu/~geology/geol101/sedimentary/seds.htm)

Obtaining Resources

The inquiry focus of *Investigating Earth Systems* will require teachers to obtain local or regional maps, rocks, and data. The *Investigating Earth Systems* web site helps teachers to find such materials. The web page for each *Investigating Earth Systems* module provides a list of relevant web sites, maps, videos, books, and magazines. Specific links to sources of these materials are often provided.

Managing Inquiry in Your Investigating Earth Systems Classroom

Materials

The proper management of materials can make the difference between a productive, positive investigation and a frustrating one. If your school has purchased the materials kit (available through It's About Time Publishing) most materials have been supplied. In many cases there will be additional items that you will need to supply as well. This can include photocopies or transparencies (**Blackline Masters** are available in the **Appendix**), or basic classroom supplies like an overhead projector or water source. On occasion, students will bring in materials. If you do not have the materials kit, a master list of materials for the entire module precedes the first investigation. Tips on using and managing materials accompany each investigation.

Safety

Each activity has icons noting safety concerns. In most cases, a well-managed classroom is the best preventive measure for avoiding danger and injury. Take time to explain your expectations before beginning the first investigation. Read through the investigations with your students and note any safety concerns. The activities were designed with safety in mind and have been tested in classrooms. Nevertheless, be alert and observant at all times. Often, the difference between an accident and a calamity is simple monitoring.

Time

This module can be completed in six weeks if you teach science in daily 45-minute class periods. However, there are many opportunities to extend the investigations, and perhaps to shorten others. The nature of the investigations allows for some flexibility.

An inquiry approach to science education requires the careful management of time for students to fully develop their investigative experience and skills. To help you manage time, each activity in the module comes with a matrix that breaks activities down into parts. It is designed to help you think about what you might accomplish with your students in 40 minutes of working class time, and what you might consider assigning for homework.

Most investigations will not easily fit into one 45-minute lesson. You may feel it necessary to extend them over two or more class periods. Some investigations include long-term studies. Where this is the case you may need to allow time for data collection each day, even after moving on to the next investigation.

Classroom Space

On days when students work as groups, arrange your classroom furniture into small group areas. You may want to have two desk arrangements—one for group work and one for direct instruction or quiet work time.

The Student Journal

The student journal is an important component of each *IES* module. (See the **Appendix** in this Teacher's Edition for a **Blackline Master** of the Journal cover sheet.) Your students are young scientists as they investigate Earth science questions. Encourage your students to think of themselves in this way and to see their journals as records of their investigations.

The journal serves other functions as well. It is a key component in performance assessment, both formative and summative. (Formative evaluation involves the ongoing evaluation of students' level of understanding and their development of skills and attitudes. Summative evaluation is designed to determine the extent to which instructional objectives have been achieved for a topic.) Encourage your students to record observations, data, and experimental results in their journals. Answers to **Review and Reflect** questions at the end of each investigation should also be recorded in the journal. It is very important that students have enough time to review, reflect, and update their journals at the end of each investigation.

Frequent feedback is essential if students are to maintain good journals. This is difficult but not impossible. For many teachers, the prospect of collecting and grading anywhere from 20 to over 100 journals in a planning period, then returning them the next day, seems prohibitive. This does not need to be the case. If you use a simple rubric, and collect journals often, it is possible to grade 100 journals in an hour. It may not be necessary to write comments every time you collect journals; in some cases, it is equally effective to address trends in student work in front of the whole class. For example, students will inevitably turn in journals that contain no dates and/or headings. This leaves many questions unanswered and makes their work very hard to interpret. There is no need to write this comment over and over again! You might want to consider keeping your own teacher journal for this module. This makes a great template for evaluating student journals. In addition to documenting class activities, you might want to make notes on classroom management strategies, materials and supplies, and procedural modifications. Sample rubrics are included in the **Appendix**.

Student Collaboration

The National Science Education Standards and Benchmarks for Science Literacy emphasize the importance of student collaboration. Scientists and others frequently work in teams to investigate questions and solve problems. There are times, however, when it is important to work alone. You may have students who are more comfortable working this way. Traditionally, the competitive nature of school curricula has emphasized individual effort through grading, "honors" classes, and so on. Many parents will have been through this experience themselves as students and will be looking for comparisons between their children's performance and other students. Managing collaborative groups may therefore present some initial problems, especially if you have not organized your class in this way before.

Below are some key points to remember as you develop a group approach.

- Explain to students that they are going to work together. Explain *why* ("two heads are better than one" may be a cliché—but it is still relevant).
- Stress the responsibility each group member has to the others in the group.
- Choose student groups carefully to ensure each group has a balance of ability, special talents, gender and ethnicity.
- Make it clear that groups are not permanent and they may change occasionally.
- Help students see the benefits of learning with and from each other.
- Ensure that there are some opportunities for students to work alone (certain activities, writing for example, are more efficiently done in solitude).

Student Discussion

Encourage all students to participate in class discussions. Typically, several students dominate discussion while others hesitate to volunteer comments. Encourage active participation by explicitly stating that you value all students' comments. Reinforce this by not rejecting answers that appear wrong. Ask students to clarify contentious comments. If you ask for students' opinions, be prepared to accept them uncritically.

Assessing Student Learning in Investigating Earth Systems

The completion of the final investigation serves as the primary source of summative assessment information. Traditional assessment strategies often give too much attention to the memorization of terms or the recall of information. As a result, they often fall short of providing information about students' ability to think and reason critically and apply information that they have learned. In *Investigating Earth Systems*, the solutions students provide to the final investigation in each module provide information used to assess thinking, reasoning, and problem-solving skills that are essential to life-long learning.

Assessment is one of the key areas that teachers need to be familiar with and understand when trying to envision implementing *Investigating Earth Systems*. In any curriculum model, the mode of instruction and the mode of assessment are connected. In the best scheme, instruction and assessment are aligned in both content and process. However, to the extent that one becomes an impediment to reform of the other, they can also be uncoupled. *Investigating Earth Systems* uses multiple assessment formats. Some are consistent with reform movements in science education that *Investigating Earth Systems* is designed to promote. **Project-based assessment**, for example, is built into every *Investigating Earth Systems* culminating investigation. At the same time, the developers acknowledge the need to support teachers whose classroom context does not allow them to depart completely from "traditional" assessment formats, such as paper and pencil tests.

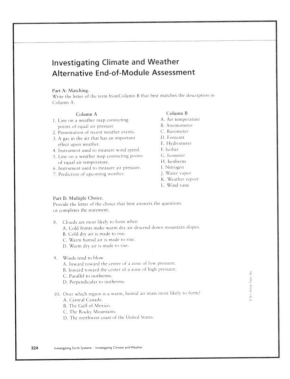

In keeping with the discussion of assessment outlined in the National Science Education Standards (NSES), teachers must be careful while developing the specific expectations for each module. Four issues are of particular importance in that they may present somewhat new considerations for teachers and students. These four issues are dealt with on the next two pages.

1. Integrative Thinking

The National Science Education Standards (NSES) state: "Assessments must be consistent with the decisions they are designed to inform." This means that as a prerequisite to establishing expectations, teachers should consider the use of assessment information. In *Investigating Earth Systems*, students often must be able to articulate the connection between Earth science concepts and their own community. This means that they have to integrate traditional Earth science content with knowledge of their surroundings. It is likely that this kind of integration will be new to students, and that they will require some practice at accomplishing it. Assessment in one module can inform how the next module is approached so that the ability to apply Earth science concepts to local situations is enhanced on an ongoing basis.

2. Importance

An explicit focus of NSES is to promote a shift to deeper instruction on a smaller set of core science concepts and principles. Assessment can support or undermine that intent. It can support it by raising the priority of in-depth treatment of concepts, such as students evaluating the relevance of core concepts to their communities. Assessment can undermine a deep treatment of concepts by encouraging students to parrot back large bodies of knowledge-level facts that are not related to any specific context in particular. In short, by focusing on a few concepts and principles, deemed to be of particularly fundamental importance, assessment can help to overcome a bias toward superficial learning. For example, assessment of terminology that emphasizes deeper understanding of science is that which focuses on the use of terminology as a tool for communicating important ideas. Knowledge of terminology is not an end in itself. Teachers must be watchful that the focus remains on terminology in use, rather than on rote recall of definitions. This is an area that some students will find unusual if their prior science instruction has led them to rely largely on memorization skills for success.

3. Flexibility

Students differ in many ways. Assessment that calls on students to give thoughtful responses must allow for those differences. Some students will find the open-ended character of the *Investigating Earth Systems* module reports disquieting. They may ask many questions to try to find out exactly what the finished product should look like. Teachers will have to give a consistent and repeated message to those students, expressed in many different ways, that the ambiguity inherent in the open-ended character of the assessments is an opportunity for students to show what they know in a way that makes sense to them. This also allows for the assessments to be adapted to students with differing abilities and proficiencies.

Teacher's Edition

4. Consistency

While the module reports are intended to be flexible, they are also intended to be consistent with the manner in which instruction happens, and the kinds of inferences that are going to be made about students' learning on the basis of them. The *Investigating Earth Systems* design is such that students have the opportunity to learn new material in a way that places it in context. Consistent with that, the module reports also call for the new material to be expressed in context. Traditional tests are less likely to allow this kind of expression, and are more likely to be inconsistent with the manner of teaching that *Investigating Earth Systems* is designed to promote. Likewise, in that *Investigating Earth Systems* is meant to help students relate Earth Science to their community, teachers will be using the module reports as the basis for inferences regarding the students' abilities to do that. The design of the module reports is intended to facilitate such inferences.

An assessment instrument can imply but not determine its own best use. This means that *Investigating Earth Systems* teachers can inadvertently assess module reports in ways that work against integrative thinking, a focus on important ideas, flexibility in approach, and consistency between assessment and the inferences made from that assessment.

All expectations should be communicated to students. Discussing the grading criteria and creating a general rubric are critical to student success. Better still, teachers can engage students in modifying and/or creating the criteria that will be used to assess their performance. Start by sharing the sample rubric with students and holding a class discussion. Questions that can be used to focus the discussion include: Why are these criteria included? Which activities will help you to meet these expectations? How much is required? What does an "A" presentation or report look like? The criteria should be revisited throughout the completion of the module, but for now students will have a clearer understanding of the challenge and the expectations they should set for themselves.

Investigating Earth Systems Assessment Tools

Investigating Earth Systems provides you with a variety of tools that you can use to assess student progress in concept development and inquiry skills. The series of evaluation sheets and scoring rubrics provided in the back of this Teacher's Edition should be modified to suit your needs. Once you have settled on the performance levels and criteria and modified them to suit your particular needs, make the evaluation sheets available to students, preferably before they begin their first investigation. Consider photocopying a set of the sheets for each student to include in his or her journal. You can also encourage your students to develop their own rubrics. The final investigation is well-suited for such, since students will have gained valuable experience with criteria by the time they get to this point in the module. Distributing and discussing the evaluation sheets will help students to become familiar with and know the criteria and expectations for their work. If students have a complete set of the evaluation sheets, you can refer to the relevant evaluation sheet at the appropriate point within an *IES* lesson.

1. Pre-Assessment

The pre-assessment activity culminates with students putting their journals together and adding their first journal entry. It is important that this not be graded for content. Credit should be given to all students who make a reasonable attempt to complete the activity. The purpose of this pre-assessment is to provide a benchmark for comparison with later work. At the end of the module, the central questions of the pre-assessment are repeated in the section called **Back to the Beginning**.

Teacher's Edition

2. Assessing the Student Journal

As students complete each investigation, reinforce the need for all observations and data to be organized well and added to the journals. Stress the need for clarity, accurate labeling, dating, and inclusion of all pertinent information. It is important that you assess journals regularly. Students will be more likely to take their journals seriously if you respond to their work. This does not have to be particularly time-consuming. Five types of evaluation instruments for assessing journal entries are available at the back of this Teacher's Edition to help you provide prompt and effective feedback. Each one is explained in turn below.

Journal–Entry Evaluation Sheet

This sheet provides you with general guidelines for assessing student journals. Adapt this sheet so that it is appropriate for your classroom. The journal entry evaluation sheet should be given to students early in the module, discussed with students, and used to provide clear and prompt feedback.

Journal–Entry Checklist

This checklist provides you and your students with a guide for quickly checking the quality and completeness of journal entries. You can assign a value to each criterion, or assign a "+" or "-" for each category, which you can translate into points later. However you choose to do this, the point is to make it easy to respond to students' work quickly and efficiently. Lengthy comments may not be necessary. Depending on time constraints, you may not have time to write comments each time you evaluate journals. The important thing is that students get feedback—they will do better work if they see that you are monitoring their progress.

Key–Question Evaluation Sheet

This sheet will help students to learn the basic expectations for the warm-up activity. The **Key Question** is intended to reveal students' conceptions about the phenomena or processes explored in the activity. It is not intended to produce closure, so your assessment of student responses should not be driven by a concern for correctness. Instead, the evaluation sheet emphasizes that you want to see evidence of prior knowledge and that students should communicate their thinking clearly. It is unlikely that you will have time to apply this assessment every time students complete a warm-up activity, yet in order to ensure that students value committing their initial conceptions to paper and taking the warm-up seriously, you should always remind students of the criteria. When time permits, use this evaluation sheet as a spot check on the quality of their work.

Investigation Journal–Entry Evaluation Sheet

This sheet will help students to learn the basic expectations for journal entries that feature the write-up of investigations. *IES* investigations are intended to help students to develop content understanding and inquiry abilities. This evaluation sheet provides a variety of criteria that students can use to ensure that their work meets the highest possible standards and expectations. When assessing student investigations, keep in mind that the **Investigate** section of an *IES* lesson corresponds to the explore phase of the learning cycle (engage, explore, apply, evaluate) in which students explore their conceptions of phenomena through hands-on activity. Using and discussing the evaluation sheet will help your students to internalize the criteria for their performance. You can further encourage students to internalize the criteria by making the criteria part of your "assessment conversations" with them as you circulate around the classroom and discuss student work. For example, while students are working, you can ask them criteria-driven questions such as: "Is your work thorough and complete? Are all of you participating in the activity? Do you each have a role to play in solving the problem?" and so on.

Review and Reflect Journal–Entry Evaluation Sheet

Reviewing and reflecting upon one's work is an important part of scientific inquiry and is also important to learning science. Depending upon whether you have students complete the work individually or within a group, the **Review and Reflect** portion of each investigation can be used to provide information about individual or collective understandings about the concepts and inquiry processes explored in the investigation. Whatever choice you make, this evaluation sheet provides you with a few general criteria for assessing content and thoroughness of student work. Adapt and modify the sheet to meet your needs. Consider involving students in selecting and modifying the criteria for evaluating their end of investigation reflections.

3. Assessing Group Participation

One of the challenges to assessing students who work in collaborative teams is assessing group participation. Students need to know that each group member must pull his or her weight. As a component of a complete assessment system, especially in a collaborative learning environment, it is often helpful to engage students in a self-assessment of their participation in a group. Knowing that their contributions to the group will be evaluated provides an additional motivational tool to keep students constructively engaged. These evaluation forms (Group–Participation Evaluation Sheets I and II) provide students with an opportunity to assess group participation. In no case should the results of this evaluation be used as the sole source of assessment data. Rather, it is better to assign a weight to the results of this evaluation and factor it in with other sources of assessment data. If you have not done this before, you may be surprised to find how honestly students will critique their own work, often more intensely than you might do.

4. Assessing the Final Investigation

Students' work throughout the module culminates with the final investigation. To complete it, students need a working knowledge of previous activities. Because it refers back to the previous steps, the last investigation is a good review and a chance to demonstrate proficiency. For an idea on how to use the last investigation as a performance-based exam, see the section in the **Appendix**.

5. Assessing Inquiry Processes

There is an obvious difficulty in assessing individual student proficiency when the students work within a collaborative group. One way to do this is to have a group present its results followed by a question-and-answer session. You can direct questions to individual students as a way of checking proficiency. Another is to have every student write a report on his or her role in the investigation, after first making it clear what this report should contain. Individual interviews are clearly the best option but may not be feasible given the time constraints of most classes.

6. Traditional Assessment Options

A traditional paper-and-pencil exam is included in the **Appendices**. While performance-based assessments may offer teachers more insight into student skill levels, computer-generated tests are also useful—especially so in states with state-sponsored exams. Additionally, some students are strong in one area and not as strong in another. Using a variety of methods for assessing and grading students' progress offers a more complete picture of the success of the student—and the teacher.

Reviewing and Reflecting upon Your Teaching

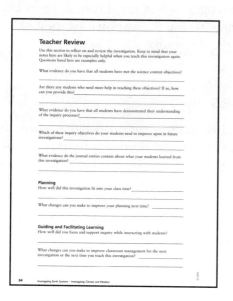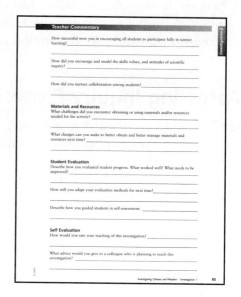

Reviewing and Reflecting upon Your Teaching provides an important opportunity for professional growth. A two-page Teacher Review form is included at the end of each investigation. The purpose of these reviews is to help you to reflect on your teaching of each investigation. We suggest that you try to answer each question at the completion of each investigation, then go back to the relevant section of this Teacher's Edition and write specific comments in the margins. Use the comments the next time you teach the investigation. For example, if you found that you were able to make substitutions to the list of materials needed, write a note about those changes in the margin of that page of this Teacher's Edition.

Teacher's Edition

GETIT™ Geoscience Education Through Interactive Technology for Grades 6-12

Earthquakes, volcanoes, hurricanes, and plate tectonics are all subjects that deal with energy transfer at or below the Earth's surface. The GETIT CD-ROM uses these events to teach the fundamentals of the Earth's dynamism. GETIT contains 63 interactive—inquiry-based—activities that closely simulate real-life science practice. Students work with real data and are encouraged to make their own discoveries—often learning from their mistakes. They use an electronic notebook to answer questions and record ideas, and teachers can monitor their progress using the integrated class-management module. The Teacher's Guide includes Assessments, Evaluation Criteria, Scientific Content, Graphs, Diagrams and Blackline Masters. GETIT conforms to the National Science Education Standards and the American Association for the Advancement of Science benchmarks for Earth Science.

Enhancing *Investigating Climate and Weather* with GETIT

Investigating Climate and Weather				
Investigation	Key Question	Page	Goals	GETIT Activity
1. Observing Weather	How is weather observed?	C8	Air Temperature	• Temperature and heat
		C8	Clouds	• Relative humidity
5. The Causes of Weather		C52	Temperature and air pressure	• Does temperature influence volume? • Does pressure influence volume? • Does volume influence density?
		C55	Evaporation and condensation	• Now you see it, now you don't • Science Showtime episode: Shake your molecules • Relative humidity?
6. Climates		C62	Ability of different materials to hold heat	• Temperature and heat
7. Exploring Climate Change	What evidence suggests that climate has changed in the past?	C69		• Tropical storm trends in North America

NOTES

Investigating Climate and Weather: Introduction

Few everyday phenomena are as fascinating as the weather. Humankind's desire to know what tomorrow's weather will be has existed for millennia, but only in about the last century have meteorologists developed scientifically based techniques for predicting the weather—and we all know that weather forecasting is still an uncertain endeavor. One of the most fundamental questions one might ask about the weather is why, in most regions of the world, the weather is so variable, when the sun's delivery of solar energy to the Earth, and the Earth's seasonal progression around the Sun, are so regular.

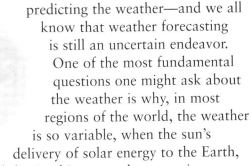

The Earth's atmosphere, a mixture of several gases, is a very thin envelope around the Earth. It has no well-defined top—its density tails off gradually upward—but most of it lies below an altitude of only several kilometers, a tiny fraction of the radius of the Earth. Just as with the world's oceans, the atmosphere is absolutely thick but relatively thin. A photo of the sunlit Earth from space shows the cloud-dotted atmosphere to be no more than a thin film above the Earth's surface.

The atmosphere is a giant heat engine. On average, the Earth's surface receives much more solar energy in low-latitude regions than at high latitudes. That imbalance necessitates a constant transport of heat from low latitudes to high latitudes, because over the long term, average temperatures in all regions of the Earth are very close to being constant. Together with ocean currents, the wind patterns of the atmosphere carry out this transport of heat, in the form of a giant and extremely complex convection system. Much of the complexity of the Earth's winds comes about because of the Earth's rotation: when a moving object is viewed by an observer on a rotating surface, the object seems to be deflected from a straight-line path. This effect, called the Coriolis effect, has profound implications for both winds and ocean currents on the Earth. A glance at a surface weather map shows immediately that winds blow almost parallel to isobars (curves of equal atmospheric pressure) rather than directly from areas of high pressure to areas of low pressure, which is what an uninitiated observer might expect.

Water vapor constitutes only a small percentage of the atmosphere, but it is the only atmospheric gas that undergoes changes in state among vapor, liquid, and solid. Evaporation at the Earth's surface adds water vapor to the atmosphere; at certain times and places, condensation of the water vapor leads to clouds and precipitation. Condensation and precipitation come about mainly by the rise of moist air. As air rises, it expands, because of the decrease in pressure upward in the atmosphere. As it expands, it cools, because it does work on its surroundings as it expands. In many situations, the cooling is sufficient to cause condensation of some of the water vapor, because the maximum water-vapor concentration in a parcel of air decreases as the temperature of the air decreases. Given that general effect, the details of when and where precipitation develops, whether locally in the form of thunderstorms or regionally in the form of large storm systems, depends in complicated ways on the nature of the circulation of the atmosphere.

Climate can be thought of as the long-term average of weather in a given region of the Earth. Climate is more than just the average of weather, however, because it involves the nature of extremes in temperature and precipitation around the long-term average. Many elements of climate have significance in determining the climate of a region: average temperature and precipitation, spread of temperature and precipitation around the mean, and the timing of temperature and precipitation from season to season. Climatologists have developed a variety of classifications of climate on the basis of these elements. The single characteristic that is most sensitive to climate is vegetation.

Long-term climate change is much in the news nowadays, because the evidence is strong that the average surface temperature of the Earth is slowly increasing. There continues to be debate about how much of that temperature increase is natural and how much is induced by the burning of fossil fuels and by deforestation. It is generally believed by climatologists, although not universally accepted, that much of the temperature increase is being caused by the great increase in atmospheric carbon dioxide brought about by human activities. Carbon dioxide is one of several atmospheric gases that are called greenhouse gases, because they are transparent to incoming short-wave solar radiation but intercept some of the outgoing long-wave radiation from the Earth's surface to outer space and reradiate the heat back to the surface. In that way, they alter the heat balance in the atmosphere so as to make the average temperature of the Earth's surface higher.

Climatologists have learned much about past climates by studying various proxies for temperature. A proxy is some measurable feature that is a function of temperature. If the relationship between the observed feature and the temperature is known, then the proxy can be used to evaluate past temperature. In recent years, bubbles of the atmosphere trapped deep in glacier ice have been a rich source of information on past atmospheric temperatures for the past 400,000 years. The chain of reasoning is not simple, however: the effect comes about by slight fractionation of the two main isotopes of the element oxygen when water evaporates and condenses, and how that fractionation varies with temperature.

Introduction

More Information…on the Web

Go to the *Investigating Earth Systems* web site www.agiweb.org/ies for links to a variety of other web sites that will help you deepen your understanding of content and prepare you to teach this module, *Investigating Climate and Weather*.

Students' Conceptions about Climate and Weather

Most students will have become aware of weather very early in their lives. They will have associated it with weather-dependent activities, weather reports, significant seasonal events, and decisions about which clothes they should wear on any given day. They know that weather plays an important role in their lives. However, they may have little understanding of how the various elements of weather combine to form complex systems. For example, they may not understand how clouds form, or how passing weather fronts can produce powerful thunderstorms. They may also not truly understand how or why scientists study weather, or how the weather reports they see on television or in the newspaper are generated.

While students are likely to have some understanding of climate, they may see it as synonymous with weather and have very little idea of the basis of climate as weather patterns over very long periods of time.

It is crucial that you find out what informal ideas your students already have about climate and weather before beginning on this module. The pre-assessment activity will tell you much of what you need to know in addressing your students' unique needs.

Investigating Climate and Weather: Module Flow

Activity Summaries	Emphasis
Pre-Assessment Students describe their understanding of key concepts explored in the module.	Recording initial content knowledge and understanding.
Introducing Climate and Weather Students discuss their ideas and experiences related to the topics they will be investigating.	Putting the investigations into a meaningful context.
Investigation 1: Observing Weather Students learn how scientists study weather by designing their own method for making weather observations. Students then plot class weather observations on a map.	Devising and carrying out a plan, constructing and using models, recording observations, synthesizing results, and sharing findings.
Investigation 2: Comparing Weather Reports Students compare weather reports from a variety of sources, evaluating them for the information they contain as well as for accuracy.	Recognizing words and phrases that relate to weather. Using a range of reports as data sources.
Investigation 3: Weather Maps Students become familiar with the information found on weather maps. They complete a simple experiment to help them understand how warm and cold air masses interact. Students also investigate how atmospheric pressure changes with altitude.	Using tools for investigation, constructing and using models, recording observations, and recognizing map symbols and conventions.
Investigation 4: Weather Radiosondes, Satellites, and Radar Students plot data from radiosondes to understand how temperature changes with altitude. Students study the relationship between weather maps, satellite images, and radar images.	Using tools for investigation, recognizing map symbols and conventions, interpreting findings, and sharing findings.
Investigation 5: The Causes of Weather Students use models to understand the factors that influence weather, including the effects of wind, cloud formation, and temperature and air pressure.	Investigating properties using models and experiments, interpreting findings, and sharing findings.
Investigation 6: Climates Students investigate the factors that define and affect climate.	Forming questions for inquiry, piloting designs, developing models, recording observations, analyzing results, and sharing findings.
Investigation 7: Exploring Climate Change Students investigate possible climate changes in the recent and distant past.	Forming questions for inquiry, examining evidence, recording observations, and sharing findings.
Investigation 8: Climate Change Today Students use their knowledge of climate and weather to make predictions about future climates.	Synthesizing results, applying results of previous experiments, and communicating findings.
Reflecting Students review the science content and inquiry processes they used throughout the module.	Assessing student learning.

Introduction

Investigating Climate and Weather: Module Objectives

Investigation	Science Content	Inquiry Process Skills
Investigation 1: Observing Weather Students learn how scientists study weather by designing their own method for making weather observations. Students then plot class weather observations on a map.	Students will collect evidence that: 1. Different aspects of the weather can be measured using specially designed instruments. 2. Air temperature is a measure of the average speed of the motion of gas molecules in the atmosphere. The greater the energy of motion of the molecules, the higher the temperature of the air. 3. Clouds are formed when humid air rises upward. As the air rises, it expands. As it expands, it becomes cooler. With enough cooling, water vapor in the air condenses into tiny water droplets, which are then visible as clouds. 4. Wind blows because air pressure is higher in one place than in another place, and air moves from areas of higher pressure to areas of lower pressure. 5. Using protocols to make a measurement increases the reliability of that measurement.	Students will: 1. Follow protocols to make reliable measurements. 2. Develop protocols that fit a format. 3. Critique other protocols. 4. Communicate procedures and other information clearly. 5. Record observations and measurements. 6. Devise an organized way to record data. 7. Search for patterns in data. 8. Synthesize data into a weather report.
Investigation 2: Comparing Weather Reports Students compare weather reports from a variety of sources, evaluating them for the information they contain as well as for accuracy.	Students will collect evidence that: 1. Certain weather words are used to describe the quality of weather measurements. 2. Weather reports are descriptions of weather conditions; weather forecasts are predictions of the weather, over both the short term and the longer term. 3. Weather reports vary, depending upon their sources of information, audience, and method of transmittal. 4. Greater understanding of the science of weather, along with improvements and innovations in technology, have allowed increasingly accurate weather forecasts. However, weather prediction will never be an exact science.	Students will: 1. Predict the accuracy of different weather reports. 2. Compare weather reports from different sources. 3. Record observations. 4. Evaluate weather reports for level of information and accuracy. 5. Recognize patterns and relationships in weather reports. 6. Use data to support or refute predictions. 7. Conduct research on weather terms. 8. Arrive at conclusions. 9. Communicate observations and findings to others.

Investigating Climate and Weather: Module Objectives

Investigation	Science Content	Inquiry Process Skills
Investigation 3: Weather Maps Students become familiar with the information found on weather maps. They complete a simple experiment to help them understand how warm and cold air masses interact. Students also investigate how atmospheric pressure changes with altitude.	Students will collect evidence that: 1. A great deal of information on weather maps is embodied in symbols. 2. Weather systems typically move across the United States from west to east. 3. Warmer air masses rise over cooler air masses. 4. Air pressure decreases upward in the atmosphere because at higher levels in the atmosphere there is less air above, and therefore less weight of a unit column of air above the given level. 5. An isotherm is a curving line that connects points on the map where temperatures are the same. An isobar is a line that connects points of equal pressure.	Students will: 1. Observe and record their impressions of a phenomenon. 2. Generate questions to answer by inquiry. 3. Devise methods of answering questions using models. 4. Collect evidence from the models. 5. Analyze evidence from the models. 6. Arrive at conclusions based on evidence.
Investigation 4: Weather Radiosondes, Satellites, and Radar Students plot data from radiosondes to understand how temperature changes with altitude. Students study the relationship between weather maps, satellite images, and radar images.	Students will collect evidence that: 1. Satellite images and radar images, as well as other sources of information, are used to make weather maps. 2. Radar is used to detect the intensity of precipitation, and to tell frozen forms from unfrozen forms of precipitation. 3. Weather satellites are valuable tools for monitoring changes in the Earth system, such as the development and scale of weather systems. 4. Air temperature usually decreases with altitude.	Students will: 1. Construct a graph from a set of data. 2. Use the data and graph to investigate relationships. 3. Search for patterns and relationships using different representations of weather (maps and satellite images). 4. Arrive at conclusions based on evidence.
Investigation 5: The Causes of Weather Students use models to understand the factors that influence weather, including the effects of wind, cloud formation, and temperature and air pressure.	Students will collect evidence that: 1. The water cycle is the system of movement of water along a variety of pathways on the Earth's surface, in the Earth's oceans, and in the Earth's atmosphere. 2. Energy and water interact in the water cycle. 3. Evaporation is the process by which a substance passes from the liquid state to the vapor state. 4. Condensation—the opposite of evaporation—is the process by which a substance passes from the vapor state to the liquid state. 5. The atmosphere exerts pressure on surfaces.	Students will: 1. Ask questions about the factors that influence weather. 2. Make predictions about the questions. 3. Use models to answer inquiry questions. 4. Collect data from models. 5. Analyze data from models. 6. Arrive at conclusions based on data. 7. Share findings and conclusions with others.

Introduction

Investigating Climate and Weather: Module Objectives

Investigation	Science Content	Inquiry Process Skills
Investigation 6: Climates Students investigate the factors that define and affect climate.	Students will collect evidence that: 1. Climate is characterized by precipitation and air temperature. 2. Climate is affected by elevation, latitude, and proximity to water and to mountain ranges. 3. Different forms of matter retain heat to different extents.	Students will: 1. Use maps to investigate questions about climate. 2. Compare and contrast climatic regions around the world. 3. Experiment with the way different states of matter (solid, liquid, and gas) retain heat. 4. Collect data from experiments. 5. Arrive at conclusions based on data analysis. 6. Communicate findings and results to others.
Investigation 7: Exploring Climate Change Students investigate how climate has changed in the recent and distant past.	Students will collect evidence that: 1. Climate can and has changed over time. 2. Fossils, ice cores, tree rings, and ocean-bottom cores can provide evidence of past climates. 3. Climate can be affected both by events that originate within the Earth system and by the Earth's interactions with other planets in the solar system.	Students will: 1. Analyze climate information, over both the short term and the long term. 2. Draw conclusions about climatic conditions from a variety of sources. 3. Share findings with others.
Investigation 8: Climate Change Today Students use their knowledge of climate and weather to make predictions about future climates.	Students will collect evidence that: 1. Climate changes over time. 2. A great deal of evidence about climate must be examined to arrive at any conclusions about the direction of climate change. 3. Climate trends may be natural, or affected by human activity.	Students will: 1. Make a prediction about the direction of climate change in the future. 2. Collect evidence about climate change over as long a period of time as possible. 3. Analyze the evidence they have collected on climate change. 4. Arrive at conclusions about climate predictions based upon evidence. 5. Communicate what they have learned about climate change to others.

National Science Education Content Standards

Investigating Earth Systems is a Standards-driven curriculum. That is, the scope and sequence of the series is derived from, and driven by, the National Science Education Standards (NSES) and the American Association for the Advancement of Science (AAAS) Benchmarks for Science Literacy (BSL). Both specify content standards that students should know by the completion of eighth grade.

Unifying Concepts and Processes
- Systems, order, and organization
- Evidence, models, and explanation
- Constancy, change, and measurement
- Evolution and equilibrium

Science as Inquiry
- Identify questions that can be answered through scientific investigations
- Design and conduct a scientific investigation
- Use tools and techniques to gather, analyze, and interpret data
- Develop descriptions, explanations, predictions, and models based upon evidence
- Think critically and logically to make the relationships between evidence and explanation
- Recognize and analyze alternative explanations and predictions
- Communicate scientific procedures and explanations
- Use mathematics in all aspects of scientific inquiry
- Understand scientific inquiry

Physical Science
- Properties and changes of properties in matter

Life Science
- Populations and ecosystems

Earth and Space Science
- Structure of the Earth system
- Earth's history
- Earth in the Solar System

Science and Technology
- Abilities of technological design
- Understandings about science and technology

Science in Personal and Social Perspectives
- Personal health
- Populations, resources, and environment
- Science and technology in society

History and Nature of Science
- Science as a human endeavor
- Nature of science

Introduction

Key NSES Earth Science Standards Addressed in IES Climate and Weather

1. Water, which covers the greater part of the Earth's surface, circulates through the crust, oceans, and atmosphere in what is known as the "water cycle." Water evaporates from the Earth's surface, rises and cools as it moves to higher altitudes, condenses as rain or snow, and falls to the surface. There, it collects in lakes, oceans, soil, and in rocks underground.
2. Clouds, formed by the condensation of water vapor, affect weather and climate.
3. Global patterns of atmospheric movement influence local weather. Oceans have a major effect on climate, because water in the oceans holds a large amount of heat.

Key AAAS Earth Science Benchmarks Addressed in IES Climate and Weather

The Physical Setting, Section B: The Earth

1. Because the Earth turns daily on an axis that is tilted relative to the plane of the Earth's yearly orbit around the Sun, sunlight falls more intensely on different parts of the Earth during the year. The difference in heating of the Earth's surface produces the planet's seasons and weather patterns.
2. Climates have sometimes changed abruptly in the past as a result of changes in the Earth's crust, such as volcanic eruptions or impacts of huge rocks from space. Even relatively small changes in atmospheric or ocean content can have widespread effects on climate if the change lasts long enough.
3. The cycling of water in and out of the atmosphere plays an important role in determining climatic patterns. Water evaporates from the surface of the Earth, rises and cools, condenses into rain or snow, and falls again to the surface. The water falling on land collects in rivers and lakes, soil, and porous layers of rock, and much of it flows back into the ocean.
4. Heat energy carried by ocean currents has a strong influence on climate around the world.

Materials and Equipment List for Investigating Climate and Weather

Pre-Assessment
Each group of students will need:
- poster board, poster paper, or butcher paper
- student journal cover sheet, one for each student (**Blackline Master** *Climate and Weather* P.2, available at the back of this Teacher's Edition)

Teachers will need:
- overhead projector, blackboard, or flip-chart paper
- transparency of **Blackline Master** *Climate and Weather* P.1 (Questions about Climate and Weather) available at the back of this Teacher's Edition

Materials Needed for Each Group per Investigation

Investigation 1
Part A
- reference resources about weather (books CD-ROMs, Internet access)
- weather instruments (thermometers, wind vane, anemometer, rain gauge, etc.)
- instructions for using each weather instrument
- weather observation sheets
- graph paper for making charts

Part B
- weather instruments (thermometers, wind vane, anemometer, rain gauge, etc.)
- instructions for using each weather instrument
- weather observation sheets
- street map of town, city, county, or school district (5 copies for the class)*
- white, self-adhesive labels (about 2.5 cm x 2.5 cm)

Investigation 2
- sample weather report/forecast
- weather reports/forecasts for three consecutive days from the following media: television, national newspaper, local newspaper, commercial or public radio, the NOAA weather radio, Internet, telephone
- weather data for the same three consecutive days as used for the weather forecasts (temperature, wind speed, precipitation, cloud cover, humidity, etc.)
- reference resources about weather

* Visit the *Investigating Earth Systems* web site www.agiweb.org/ies for suggestions on obtaining these resources.

Introduction

Investigation 3

Part A
- 3 weather maps from different newspapers
- information about weather-map symbols
- weather map with symbols for cloudy skies, partly cloudy skies, and rain*
- weather map with symbols for high-pressure and low-pressure areas*
- overhead transparency sheet
- 2 or 3 colors of overhead transparency markers
- weather map with temperatures only*
- weather map showing 10° isotherms*
- colored pencils

Part B
- two identical clear plastic 500-mL bottles with medium-sized necks
- caps for the bottles
- one piece of poster board, 10 cm x 10 cm
- supply of hot and cold water
- food coloring: red and blue
- sink or large pan

Part C
- aneroid barometer with scale marked in inches of mercury

Investigation 4

Part A
- two sheets of graph paper

Part B
- satellite images*
- weather maps*
- recordings of television weather broadcasts (optional)

Part C

no additional materials needed

Investigation 5

Part A
- paper towels

* Visit the *Investigating Earth Systems* web site www.agiweb.org/ies for suggestions on obtaining these resources.

Part A

Station 1
- water supply
- battery-powered fan
- two alcohol thermometers
- tape
- cotton batting
- piece of stiff cardboard

Part A

Station 2
- Styrofoam® picnic cooler
- brick or other heavy mass
- metal container, with lid, small enough to fit in cooler but large enough to contain the brick
- ice
- flashlight

Part A

Station 3
- two large round balloons (same size)
- meter stick or measuring tape (or string and meter stick)
- two thermometers
- ice bath

Part B

no additional materials needed

Investigation 6

Part A
- blank global map* or **Blackline Master** *Climate and Weather* 6.2
- colored pencils or markers

Part B
- three heat-resistant containers with a pencil-sized hole punched in the center of each lid
- water supply
- sand

* Visit the *Investigating Earth Systems* web site www.agiweb.org/ies for suggestions on obtaining these resources.

Introduction

- three thermometers
- heat lamp
- graph paper

Part C
- climate resources (books, CD-ROMs, Internet access, etc.)
- poster board and presentation supplies

Investigation 7
- resources on global climate change (books, CD-ROMs, Internet access, etc.)

Investigation 8
- students' data from previous investigations
- weather data for community for past 30 years*
- resources on global climate change (books, CD-ROMs, Internet access, etc.)
- presentation materials (poster board, markers, etc.)

General Supplies
Although the investigations can be done with the specific materials listed, it is always a good idea to build up a supply of general materials.

- 2 or 3 large clear plastic storage bins about 30 cm x 45 cm x 30 cm deep, with lids (these can be used for storage and also make good water containers)
- 2 or 3 plastic buckets and one large water container (camping type with a faucet)
- rolls of masking tape, duct tape, and clear adhesive tape
- rolls of plastic wrap and aluminum foil
- clear self-locking plastic bags (various sizes)
- ball of string and spools of sewing thread
- pieces of wire (can be pieces of wire coat hangers)
- stapler, staples, paper clips, and binder fasteners

- safety scissors and one sharp knife
- cotton balls, tongue depressors
- plastic and paper cups of various sizes
- empty coffee and soup cans, empty boxes and egg cartons
- several clear plastic soda bottles (various sizes)
- poster board, overhead transparencies, tracing paper, and graph paper
- balances and/or scales, weights, spring scales
- graduated cylinders, hot plates, and microscopes
- safety goggles
- disposable latex gloves
- lab aprons or old shirts
- first aid kit

* Visit the *Investigating Earth Systems* web site www.agiweb.org/ies for suggestions on obtaining these resources.

Pre-assessment

Overview

During the pre-assessment phase, the students complete an open-ended survey of their knowledge and understanding of key concepts explored in the *Climate and Weather* module. Students are given four questions to consider, and their responses become the first entry in their journal.

Preparation and Materials Needed

This pre-assessment activity does not appear in the student book. However, to find out what your students already know about climate and weather, it is crucial that you conduct this pre-assessment before introducing the module and distributing the student books. When you complete the pre-assessment activity, you will have important data that tell you what your students already know about climate and weather.

Make sure that your students understand clearly that the pre-assessment is not a test. Explain that they can review the pre-assessment at the end of the module to see how their ideas and knowledge about climate and weather have changed as a result of their investigations. Tell them that the pre-assessment will also help you gauge how successful the investigations have been for everyone.

After the pre-assessment and before distributing the student books, take some time to reflect on the ideas your students have. This is the starting point. You need to ensure that what follows fits with your students' prior knowledge.

Materials:
- poster board, poster paper, or butcher paper
- overhead projector, blackboard, or flipchart paper
- overhead transparency of questions (**Blackline Master** *Climate and Weather* **P.1**)
- Student Journal cover sheet, one for each student (**Blackline Master** *Climate and Weather* **P.2**)

Suggested Teaching Procedure

1. Let students know that what they write in this exercise will become their first entry in a scientific journal that they will keep throughout the module. Explain that each person is going to write down all the ideas that he or she has about the answers to questions dealing with climate and weather. The reason for this is to provide them, and you, with a starting point for their investigations. Tell students that when they have finished the module, they will answer these same questions again. This will allow them, and you, to compare how their knowledge about climate and weather has changed as a result of their investigations.

Introduction

Emphasize to the students that they are to complete this exercise on their own, not by comparing notes or observations with their classmates.

2. Display the pre-assessment questions on an overhead projector, or provide each student with a copy of the questions (**Blackline Master** *Climate and Weather* P.1). Have students write responses to these questions in their journals:

   > - What information is contained in a weather report and how is this information obtained?
   > - What is the difference between climate and weather?
   > - What evidence is there that climate change has happened?
   > - What evidence is there that climate change is happening?

3. Give each student a copy of the journal cover sheet (**Blackline Master** *Climate and Weather* P.2). Direct students to insert the journal cover sheet and their pre-assessment into their journal. Explain that they now have one of the most important tools for this investigation into climate and weather: their own scientific journal.

   Allow a reasonable amount of time for all students to respond. Circulate around the classroom, prompting students to provide as much detail as possible.

Teaching Tip

What form will journals take? Using loose-leaf notebook paper in a thin three-ring binder enables students to add or remove pages easily. On the downside, loose-leaf pages are more easily lost and students must maintain a regular supply of paper. If you prefer to have students keep journals in composition notebooks or laboratory notebooks, have them trim the journal cover sheet to the appropriate size and paste it onto the first page of their notebooks.

4. Divide students into groups. Instruct the groups to discuss the following:
   - Ideas we have about climate and weather
   - Questions we have about climate and weather

   One member of the group should record his/her group's ideas and questions on a sheet of poster board, poster paper, or butcher paper.

5. Discuss student responses by having each group, in turn, report on its ideas. As groups are responding, build up two important lists (ideas and questions) for everyone to see (on a chalkboard, flipchart paper, poster board, or an overhead transparency).

6. Direct students to add these ideas and questions to their journals.

7. This completes the pre-assessment phase. Distribute copies of *IES Climate and Weather*.

Assessment Opportunity

It will be useful for you to review what your students have written before moving further into the module. This will alert you in advance to any specific problems they may encounter when beginning the module. Keep these lists. They also represent pre-assessment data, and you will be able to revisit them with your students at the end of this climate and weather module to help track changes in understanding.

Introduction

NOTES

INVESTIGATING CLIMATE AND WEATHER

The Earth System

The Earth System is a set of systems that work together in making the world we know. Four of these important systems are:

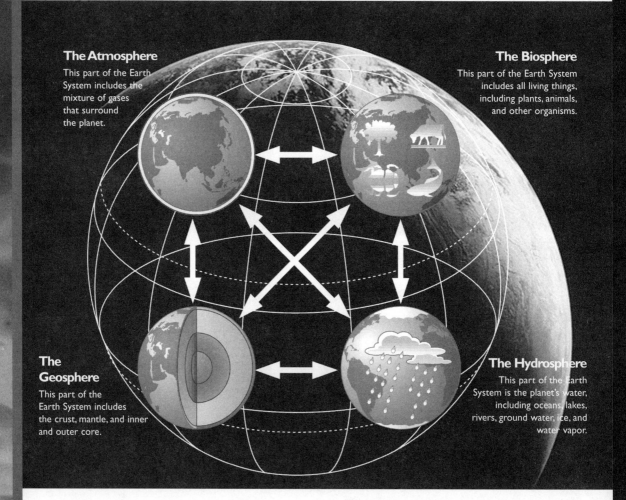

The Atmosphere
This part of the Earth System includes the mixture of gases that surround the planet.

The Biosphere
This part of the Earth System includes all living things, including plants, animals, and other organisms.

The Geosphere
This part of the Earth System includes the crust, mantle, and inner and outer core.

The Hydrosphere
This part of the Earth System is the planet's water, including oceans, lakes, rivers, ground water, ice, and water vapor.

Introduction

Introducing the Earth System

Understanding the Earth System is an overall goal of the *Investigating Earth Systems* series. The fact that Earth functions as a whole, and that all its parts operate together in meaningful ways to make the planet work as a single unit, underlies each module. Each module guides the students to consider this fundamental principle.

At the end of every investigation, students are asked to link what they have discovered with ideas about the Earth System. Questions are provided to guide their thinking, and they are asked to write their responses in their journals. They are also reminded on occasion to record the information on an *Earth System Connection* sheet. This sheet will provide a cumulative record of the connections that the students find as they work through the investigations in the module.

Not all the connections will be immediately apparent to your students. They will probably need your help to understand how some of the things they have been investigating connect to the Earth System. However, by the time they complete the *Investigating Earth Systems* modules to the end of eighth grade, they should have a working knowledge of how they and their environment function as a system within the Earth System.

The Earth System that includes the processes of climate and weather is known as the atmosphere, which includes the mixture of gases that surround the planet. Clouds and precipitation are components of weather and climate that are included in the hydrosphere. Changes in mode of operation of climate and weather on Earth affects the distribution of water within the hydrosphere. Weather and climate affect living things in the biosphere because temperature and moisture determine how successful organisms will be at survival. Changes in climate and weather also affect the rocks that make up the geosphere by their effect on erosion and deposition.

Distribute a copy of the *Earth System Connection* sheet (**Blackline Master Climate and Weather I.1**) available at the back of this Teacher's Edition. Have the students place the sheets in their journals.

Explain to the students that at the end of each investigation they will be asked to reflect on how the questions and outcomes of their investigation relate to the Earth System. Tell them that they should enter any new connections that they discover on their *Earth System Connection* sheet. Encourage them to also include connections that they have made on their own. That is, they should not limit their entries to just those suggested in the **Thinking about the Earth System** questions in **Review and Reflect**. Use the **Review and Reflect** time to direct students' attention to how local issues relate to the questions they have been investigating. By the end of the module, students should have as complete an *Earth System Connection* sheet to *Climate and Weather* as possible.

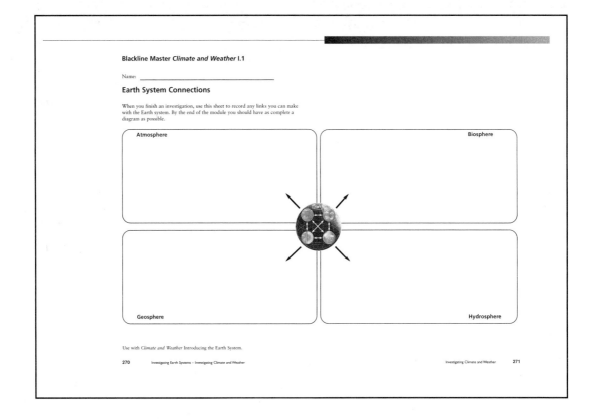

Introduction

NOTES

INVESTIGATING CLIMATE AND WEATHER

Introducing Inquiry Processes

When geologists and other scientists investigate the world, they use a set of inquiry processes. Using these processes is very important. They ensure that the research is valid and reliable. In your investigations, you will use these same processes. In this way, you will become a scientist, doing what scientists do. Understanding inquiry processes will help you to investigate questions and solve problems in an orderly way. You will also use inquiry processes in high school, in college, and in your work.

During this module, you will learn when, and how, to use these inquiry processes. Use the chart below as a reference about the inquiry processes.

Inquiry Processes:	How scientists use these processes
Explore questions to answer by inquiry	Scientists usually form a question to investigate after first looking at what is known about a scientific idea. Sometimes they predict the most likely answer to a question. They base this prediction on what they already know to be true.
Design an investigation	To make sure that the way they test ideas is fair, scientists think very carefully about the design of their investigations. They do this to make sure that the results will be valid and reliable.
Conduct an investigation	After scientists have designed an investigation, they conduct their tests. They observe what happens and record the results. Often, they repeat a test several times to ensure reliable results.
Collect and review data using tools	Scientists collect information (data) from their tests. The data may be numerical (numbers), or verbal (words). To collect and manage data, scientists use tools such as computers, calculators, tables, charts, and graphs.
Use evidence to develop ideas	Evidence is very important for scientists. Just as in a court case, it is proven evidence that counts. Scientists look at the evidence other scientists have collected, as well as the evidence they have collected themselves.
Consider evidence for explanations	Finding strong evidence does not always provide the complete answer to a scientific question. Scientists look for likely explanations by studying patterns and relationships within the evidence.
Seek alternative explanations	Sometimes, the evidence available is not clear or can be interpreted in other ways. If this is so, scientists look for different ways of explaining the evidence. This may lead to a new idea or question to investigate.
Show evidence & reasons to others	Scientists communicate their findings to other scientists to see if they agree. Other scientists may then try to repeat the investigation to validate the results.
Use mathematics for science inquiry	Scientists use mathematics in their investigations. Accurate measurement, with suitable units is very important for both collecting and analyzing data. Data often consist of numbers and calculations.

Introduction

Introducing Inquiry Processes

Inquiry is at the heart of *Investigating Earth Systems*. That is why each module title begins with the word "Investigating." In the National Science Education Content Standards, inquiry is the first content standard. (See **Science as Inquiry** on page 8 of this Teacher's Edition.)

Inquiry depends on active student participation. It is very important to remind students of the steps in the inquiry process as they perform them. Emphasize the importance of inquiry processes as they occur in your investigations. Provoke students to think about *why* these processes are important. Collecting good data, using evidence, considering alternative explanations, showing evidence to others, and using mathematics are all essential to *Investigating Earth Systems*. Use examples to demonstrate these processes whenever possible.

At the end of every investigation, students are asked to reflect on their thinking about scientific inquiry. Refer students to the lists of inquiry processes as they answer these questions.

> **Teaching Tip**
>
> If the reading level of the descriptions of inquiry processes is too advanced for some students, you could provide them with illustrations or examples of each of the processes. You may wish to provide students with a copy of the inquiry processes to include in their journals (**Blackline Master** *Climate and Weather* P.2).

Introducing Climate and Weather

Have you ever been in the middle of a powerful storm?

Have you ever wondered where the information for weather reports comes from and why weather forecasts are important?

Have you ever seen the effects of a serious lack of rain?

Have you ever seen clouds forming over a body of water?

Introduction

Introducing Climate and Weather

Use this introduction to the module to set your students' investigations into a meaningful context.

Students will have had a variety of experiences relating to climate and weather. This is an opportunity for them to offer some of their own experiences in a general discussion, using these questions as prompts. Some students may be able to cite experiences additional to those asked for here. Encourage a wide-ranging discussion.

Because your students have just spent time in the pre-assessment phase thinking about and discussing what they already know about weather, it probably is not necessary to have them complete another journal entry. They will be anxious to get to work on their investigations.

You may want to quickly summarize the main points that emerge from the discussion. You could do this on a chalkboard, an easel pad, or an overhead transparency. For your own assessment purposes, it will be useful to keep a record of these early indicators of student understanding.

> **About the Photos**
>
> The upper left photograph shows palm trees being blown by the strong winds of a hurricane. The upper right photograph shows a meteorologist at work analyzing weather data. These include satellite images with identifiable storm systems (shown as white swirling masses on the black-and-white photographs). Meteorologists are scientists who study weather; they are the scientists responsible for producing weather reports and forecasts that help us to plan our daily activities. The lower left photograph shows a dry riverbed. The lack of water has caused the mud surface to crack and peel. The lower right photograph shows cumulus clouds forming over the ocean in Florida.

INVESTIGATING CLIMATE AND WEATHER

Why Are Climate and Weather Important?

"What's the weather going to be like today?" That is often the first question on your mind when you wake up in the morning. What you wear, and sometimes even what you do on any given day depends on the weather. Weather can change very quickly. It could be sunny and warm in the morning, and in the afternoon you could be faced with dangerous thunderstorms and even tornadoes. You count on meteorologists (scientists who study the weather) to provide you with daily weather information.

On the other hand, you depend on the climate to give you fairly similar weather conditions year after year. Farmers expect the same length of growing season each year. Ski-resort operators anticipate a reasonable snowfall each year. They rely on the climate in the area to remain the same. Yet over very long periods of time, climate can change. Climatologists (scientist who study the climate) have evidence that the climate has changed many times in the past.

What Will You Investigate?

These investigations will put you in the roles of weather reporter, fact finder, and inquiring student. You will be using weather instruments and observations to make weather maps, weather reports, and weather forecasts. You will look for patterns in your weather data and explore reasons for those patterns. You will explore how climate has changed over time, the effects of climate on your life now, and what might happen if the climate changes in the future.

Here are some of the things that you will investigate:

- how weather instruments work;
- what is contained in a weather report and map;
- how weather observations are made;
- the underlying causes of weather patterns;
- the difference between climate and weather;
- how scientists know that the climate has changed in the past;
- how climate is changing now.

Introduction

Why Are Climate and Weather Important?

Read (or have a student read) this section aloud. You may want to discuss the difference between climate and weather on the basis of the information given in the student text. Most students will have some clear ideas about what weather is and how it affects them, but they are less likely to know the difference between climate and weather. Most students are likely to think that climate and weather are the same thing. From reading the text, students should understand that weather refers to a particular time, day to day, whereas climate is the long-term average of weather in a particular region of the Earth.

> **About the Photo**
> The photograph of cyclists riding during rainy conditions illustrates how weather affects our everyday lives.

What Will You Investigate?

Give students a sense of where they are headed in the module. Help them to connect what may seem like unrelated investigations into a cohesive network of ideas. Reviewing this section of the introduction is the first step toward constructing a conceptual framework of "the Big Picture" as it is explored in *Investigating Climate and Weather* (see page C88 of the student text). This framework includes five main concepts:

- how weather observations are made and used to produce weather maps and forecasts;
- what causes weather patterns;
- how climate differs from weather;
- how scientists study climate change through time;
- how climate is changing now.

This would be a good time to review with students the titles of the activities in the table of contents. Ask students to explain how these titles relate to the descriptions in **What Will You Investigate?**

Discussing the final investigation will help students to understand the overall goal of the module. In **Investigation 8**, they use all the skills and knowledge they have gained to make predictions about what the climate will be like in their community a century from now. Consider introducing students to evaluation rubrics so that they can see how their work will be assessed. Sample rubrics are included in the back of this Teacher's Edition.

NOTES

Teacher Commentary

INVESTIGATION 1: OBSERVING WEATHER

Background Information

What is Temperature?
All matter consists of atoms, which are bonded in some cases into larger particles called molecules. At all temperatures above absolute zero (about -273°C), the atoms are in constant random motion, called thermal motion. In solids, they vibrate around fixed positions; in gases, they take long, flying trajectories before colliding with the walls of a container or with other atoms or molecules in the gas. The energy associated with these motions is called thermal energy, or heat. The temperature of a body is an indirect measure of the thermal energy of a material, as compared to that of another material. Several quantitative scales of temperature have been developed. They are all derived on the basis of the following observation: If we put two bodies in contact with one another, and heat flows from the first body to the second body, then we say that the first body has a higher temperature then the second body. We sense objects as hot or cold depending on whether the temperature of our fingers is greater or less than that of the material we touch. When we touch a hot object, the faster-moving atoms of that material act to speed up the atoms of our fingers, and in the process they are slowed down. The result is a transfer of heat from the object to our fingers, as well as a decrease in the temperature of the object and an increase in the temperature of our fingers.

Atmospheric Pressure
Atmospheric pressure is the force, per unit surface area, exerted by the atmosphere on a surface of a material. The origin of the pressure is just the enormous number of collisions of the speeding molecules of the atmosphere with the surface. The total force exerted by the atmosphere on the surface is then the product of the pressure and the area of the surface.

Although somewhat less easy for your students to understand, atmospheric pressure also acts everywhere in the atmosphere itself. Think about an imaginary planar surface anywhere in the air. If there were a solid surface located at that plane, an actual pressure force would be exerted. That leads to the concept that atmospheric pressure is transmitted everywhere through the atmosphere.

Your students might wonder why they can't feel atmospheric pressure on the surfaces of their bodies, even though the pressure at sea level is about 14.7 pounds per square inch (and atmospheric pressure is substantial even at elevations high above sea level). The reason is that the internal pressure of our bodies adjusts to match the atmospheric pressure outside our bodies. If your students are able to take the elevator ride described in **Part C** of **Investigation 3**, they might feel their ears "pop" as they go up and down. That's a result of rapidly changing atmospheric pressure relative to the existing pressure within their ears.

The origin of atmospheric pressure is easy to understand. Air has weight. The weight of the column of air above any unit area of the land surface is just the atmospheric pressure at that point on the land surface. The simple reason why pressure decreases with height is that the total weight of a unit column of air above any given elevation decreases with height. It's the same as with a column of water in a tall drinking glass, with one important exception: unlike water, air is compressible, meaning that it contracts when a force of compression is exerted on it. Because of that, the density of the atmosphere decreases with elevation as well.

Wind

The fundamental reason why winds blow is that atmospheric pressure varies from place to place at a given level in the atmosphere. If the pressure is higher at one place than at another place at the same level, the air tends to move from the area of higher pressure to the area of lower pressure, unless some other opposing force counteracts that horizontal pressure force. The speed and direction of wind that results from pressure differences are more complicated than they might seem, however. The reason has to do with the effect of the Earth's rotation, which leads to some counterintuitive but very important effects. See the **Background Information** section in **Investigation 3** for more details.

Thermometers

Thermometers are certainly the most familiar weather instruments. The most common thermometers for measuring temperature at the Earth's surface are the liquid-in-glass type. These have a small glass bulb at the end of a long glass tube that is sealed at both ends. A very small tubular hollow center, called the bore, extends from the bulb to the opposite end of the tube. The tube is graduated in degrees Fahrenheit or degrees Celsius, or both. A liquid in the bulb—almost always either mercury or colored alcohol—can move up the bore. When the air temperature rises, the liquid in the bulb expands more than the glass of the thermometer, causing the liquid to rise up in the tube. Because the bore is very narrow, a small temperature change causes a large change in the length of the liquid in the tube.

Specialized liquid-in-glass thermometers measure the maximum or minimum temperature that is reached during some period of time. Maximum thermometers are the same as ordinary thermometers except that the base of the bore, where it joins the bulb, has a very narrow constriction. Expansion during rising temperature forces the liquid through the constriction, but if the thermometer is mounted horizontally, the liquid cannot pass back through the constriction when the temperature decreases again. The thermometer is reset by rapid spinning, which greatly increases the downward force on the liquid column and forces it back through the constriction. Home "fever" thermometers are just such maximum thermometers. Many of your students might have had experience in "shaking down" a fever thermometer before taking someone's temperature.

Minimum thermometers are the same as ordinary thermometers except that the liquid column in the tube contains a tiny solid object in the shape of a miniature dumbbell. When the thermometer is mounted horizontally, the dumbbell is pulled down with the shrinking column, because surface tension at the top of the column prevents it from passing into the air above the liquid column. As the temperature rises again, the dumbbell stays at the minimum-temperature position. The thermometer is reset either by spinning, to force the dumbbell back against the top of the liquid column, or, if the dumbbell is made of iron, by using a magnet to draw it back into position.

There are other kinds of thermometers as well as the liquid-in-glass type. Electrical thermometers work on the principle that the resistance of a metal wire increases with temperature. A reading of the resistance of the wire when an electrical current is passed through it is converted, by calibration, to the temperature of the wire. Bimetallic thermometers work on the principle of differential thermal expansion. They consist of two pieces of metal, of different compositions, bonded together to form a single strip. As the temperature changes, one metal expands more than the other, setting up stresses that cause the strip to bend. The bend is amplified by mechanical coupling to turn a needle relative to a calibrated scale.

Teacher Commentary

Barometers
Because atmospheric pressure is measured using barometers, it is often called barometric pressure. The barometer was invented by the Italian scientist Torricelli in the seventeenth century. The design of Torricelli's barometer is still used today. It consists of a long vertical glass tube, sealed at the top and open at the bottom. All of the air is removed from the tube, and the lower end is immersed in a cup of mercury. When the lower end is opened, the mercury rises in the tube because of the pressure of the air on the surface of the mercury in the supply cup. The greater the atmospheric pressure, the higher the mercury rises in the tube. Mercury must be used because a less dense liquid would rise impractically high; water would rise more than three meters!

The most common barometers are called aneroid barometers. They are much less expensive than mercury barometers, but they are not as accurate. At the heart of an aneroid barometer lies a small, sealed metal box in the shape of a thick disk. Small changes in the force exerted on the flat walls of the box by the atmospheric pressure cause the walls to expand or contract. This movement is amplified and transmitted to a dial by a system of levers.

Anemometers
Early anemometers consisted of a rectangular metal plate hinged along its top edge and free to swing with the wind. The metal plate is pushed from the vertical by the force of the wind. The angle relative to the vertical is calibrated to the speed of the wind. Such anemometers are simple but not very accurate.

Most modern anemometers are cup anemometers, consisting of three or four hemispherical cups mounted on horizontal shanks that radiate from a vertical shaft. The cup assembly spins in the wind, because the force of the wind on the open side of the cup is greater than that on the rounded side of the cup. The speed of rotation is calibrated to the wind speed. Because of the opposing force of the wind on the cups as they make their "return trip," the speed of movement of the cups is less than the speed of the wind.

Clouds and Precipitation
See the **Background Information** section of **Investigation 5** for details about clouds and precipitation.

More Information…on the Web
Go to the *Investigating Earth Systems* web site www.agiweb.org/ies for links to web sites that will help you deepen your understanding of content and prepare you to teach this investigation.

Investigation Overview

Students specialize in learning about one kind of weather measurement and instrument. After researching the science behind their weather observation, they devise a plan to collect data and then carry out the plan. Students then design a center to teach others about their weather observation. After making weather observations, the class compiles five daily weather maps. **Digging Deeper** describes the elements of weather, including air temperature, wind, clouds, and precipitation.

Goals and Objectives

As a result of **Investigation 1**, students will develop a better understanding of weather observations and will improve their ability to make and record observations.

Science Content Objectives

Students will collect evidence that:
1. Different aspects of the weather can be measured using specially designed instruments.
2. Air temperature is a measure of the average internal energy of motion of gas molecules in the atmosphere. The greater the internal energy of motion of the molecules, the higher the temperature of the air.
3. Clouds are formed when humid air rises upward. As the air rises, it expands. As it expands, it becomes cooler. With enough cooling, water vapor in the air condenses into tiny water droplets, which are then visible as clouds.
4. Wind blows because air pressure is higher in one place than in another place, and air moves from areas of higher pressure to areas of lower pressure.
5. Using protocols to make a measurement increases the reliability of that measurement.

Inquiry Process Skills

Students will:
1. Follow protocols to make reliable measurements.
2. Develop protocols that fit a format.
3. Critique other protocols.
4. Communicate procedures and other information clearly.
5. Record observations and measurements.
6. Devise an organized way to record data.
7. Search for patterns in data.
8. Synthesize data into a weather report.

Teacher Commentary

Connections to Standards and Benchmarks
In **Investigation 1**, students will study the instruments that are used to compile the measurements used in weather reports and weather maps. These observations will start them on the road to understanding the National Science Education Standards and AAAS Benchmark shown below.

NSES Links
- Different kinds of questions suggest different kinds of scientific investigations. Some investigations involve observing and describing objects, organisms, or events; some involve collecting specimens; some involve experiments; some involve seeking more information; some involve discovery of new objects and phenomena; and some involve making models.

- The atmosphere is a mixture of nitrogen, oxygen, and trace gases that include water vapor. The atmosphere has different characteristics at different altitudes.

AAAS Link
Because the Earth turns daily on an axis that is tilted relative to the plane of the Earth's yearly orbit around the Sun, sunlight falls more intensely on different parts of the Earth during the year. The difference in heating of the Earth's surface produces the planet's seasons and weather patterns.

Preparation and Materials Needed

Preparation

For **Part A**, you will need to collect or order weather instruments, or materials to make weather instruments. Suggestions for weather instruments that students may use include an indoor/outdoor thermometer, an anemometer, a wind vane, and a rain gauge. Instructions for making various weather instruments can be found on the *Investigating Earth Systems* web site and are also given in **Blackline Masters** *Climate and Weather* 1.2, 1.3 and 1.5. You will need to collect reference resources about weather, like books, CD-ROMs, or Internet articles. You will also need to obtain a street map of your town, city, county, or school district.

For **Part B**, students are asked to make weather observations at home for a week. You will need to think about how this will fit into your class plans. You may want students to collect all of their data and then finish the investigation, or you can have them bring in their observations each day to be added to the class weather maps. Suggestions and information that will help students in making a variety of different weather observations are given in **Blackline Masters** *Climate and Weather* 1.1 – 1.8.

If your students are not used to working in small collaborative groups, spend some time helping them understand how to do this. Keep in mind that some students may find it difficult to work in a group (some prefer to work alone). Help students understand that "collaborating" means "working together," and that collaboration is an important part of scientific inquiry. Sample rubrics for evaluating group participation are provided at the back of this Teacher's Edition (see **Assessment Tools: Group Participation Evaluation Sheet I** and **Group Participation Evaluation Sheet II**). Discussing the criteria will help to reinforce the importance of individual accountability and cooperation.

Materials

Part A
- reference resources about weather (books CD-ROMs, Internet access)
- weather-recording instruments (thermometers, rain gauge, wind vane, anemometer, etc.)
- weather observation sheets
- graph paper for making charts
- large flat-bottomed plastic pail or metal

Part B
- weather measuring instruments (thermometers, barometers, pH paper, rain gauge, wind vane, anemometer, etc.)
- weather observation sheets
- street map of town, city, county, or school district (5 copies for the class)*
- white, self-adhesive labels (about 2.5 cm x 2.5 cm)

* The *Investigating Earth Systems* web site provides suggestions for obtaining these resources.

Teacher Commentary

NOTES

Investigating Climate and Weather

Investigation 1: Observing Weather

Investigation 1:
Observing Weather

Key Question

Before you begin, first think about this key question.

How is weather observed?

Think about what you know about weather reports and weather maps. What sort of information goes into one? How is this information obtained? Make a list that combines what you know about obtaining weather information with questions that you might be able to answer in this investigation. Keep your list for review later.

Share your thinking with others in your group and with your class.

Materials Needed

For this investigation your group will need:

- reference resources about weather (books, CD-ROMs, access to the Internet)
- weather instruments (thermometer, wind vane, anemometer, rain gauge)
- instructions for using each weather instrument
- graph paper for making charts
- large flat-bottomed plastic pail or metal can
- street map of town, city, county, or school district
- white, self-adhesive labels, about 2.5 cm × 2.5 cm

Investigate

Part A: Kinds of Weather Observations

1. Your group members are going to become specialists in making a weather measurement or observation.

Teacher Commentary

Key Question

Use this question as a brief warm-up activity to elicit students' ideas about how weather observations are made. Emphasize thinking and sharing of ideas. Avoid seeking closure (i.e., the "right answer"). Closure will come through inquiry, reading the text (**Digging Deeper**), discussing the ideas (lecture), and reflecting at the end of the investigation on what was learned. Make students feel comfortable sharing their ideas by avoiding commentary on the correctness of responses.

Write the **Key Question** on the blackboard or on an overhead transparency. Have students record the question and their answers in their journals. Tell students to think about and answer the question individually. Tell them to write as much as they know and to provide as much detail as possible in their responses. Emphasize that the date and the prompt (question, heading, etc.) should be included in all journal entries.

Discuss students' ideas. Ask for a volunteer to record responses on the blackboard or on an overhead transparency. This allows you to circulate among the students, encouraging them to copy the notes in an organized way.

Student Conceptions about Weather Observations

The pre-assessment will have given you a sense of what your students already know about weather and what they do not know. Many might have definite ideas, perhaps recognizing some of the key weather elements and events. However, they may simply see different weather conditions as unrelated events and have little understanding of weather as a system. They may not recognize that weather conditions move across the surface of the Earth in a systematic way.

Most students will have had some experience with weather reports and forecasts, whether from television or newspapers. They are likely to know that conditions like temperature and precipitation are usually included in a weather report, but they may not know how scientists make these observations.

Answer for the Teacher Only

Weather maps and reports include information like temperature, precipitation, sky conditions (cloud cover), wind speed and direction, and air pressure. Scientists use instruments like thermometers, wind vanes, anemometers, and rain gauges, as well as more sophisticated equipment like satellite imagery and radar, to make weather reports and maps.

Assessment Tool

Key–Question Evaluation Sheet

Use this evaluation sheet to help students understand and internalize basic expectations for the warm-up activity. The **Key–Question Evaluation Sheet** emphasizes that you want to see evidence of prior knowledge and that students should communicate their thinking clearly. You will not likely have time to apply this assessment every time students complete a warm-up activity; yet, in order to ensure that students value committing their initial conceptions to paper and are taking the warm up seriously, you should always remind them of the criteria. When time permits, use this evaluation sheet as a spot check on the quality of their work.

Teacher Commentary

Making Connections...with History

Year	Event
400 BC	Hippocrates writes on the influence of climate on health
350 BC	Aristotle writes on weather science
300 BC	Theophrastus writes on winds
1441	World's oldest rain gauge, Chuk-u-gi, is invented
1593	Galileo invents the thermometer
1643	Torricelli invents the barometer
1661	Boyle proposes his law on gases
1664	Formal weather observations begin in Paris, France
1668	Edmund Halley draws first map of the trade winds
1714	The Fahrenheit scale is introduced
1735	George Halley's treatise on the effect of the Earth's rotation on the wind
1736	The Centigrade scale is introduced
1743	Benjamin Franklin compares weather observations in different colonies and predicts that a storm's course can be plotted
1779	Formal weather observations begin in New Haven, Connecticut
1783	The hair hygrometer is invented
1802	Lamark and Howard suggest a cloud classification scheme
1817	Humboldt draws the first map of global mean annual temperature
1825	August invents the psychrometer
1827	Dove develops the laws of storms
1831	Redfield makes the first weather map of the United States
1837	The pyrheliometer is invented
1841	Espy develops a theory on storm movement
1844	Coriolis describes the "Coriolis Effect"
1845	Berghaus makes the first world map of precipitation
1848	Dove publishes the first maps of mean monthly temperature
1862	Renou makes the first map showing mean pressure in Europe
1879	Supan publishes map of world temperature regions
1892	Beginning of systematic use of weather balloons
1902	The stratosphere is discovered
1913	The ozone layer is discovered
1918	Bjerknes develops his polar front theory
1925	Beginning of systematic data collection by aircraft
1928	Radiosondes are first used on weather balloons
1940	Investigation into the nature of jet streams
1960	The U.S. launches Tiros I, the first weather satellite

Source: Jon Erickson, *Violent Storms*, Blue Ridge Summit, PA: Tab Books, Inc., 1988, p. 118

INVESTIGATING CLIMATE AND WEATHER

Your job will be to:
- research the science behind your weather observation;
- learn about the techniques for making your observation;
- study any instrument that is needed for your observation;
- practice making your weather observation;
- write a protocol for making your weather observation;
- set up a center for your classmates to learn how to make your weather observation correctly.

 Check any design with your teacher before attempting to construct a homemade instrument.

2. Your group may be assigned one or more of the following weather observations: temperature, wind speed, wind direction, cloud types, cloud cover, precipitation type, or precipitation amount.

 Conduct Investigations

Teacher Commentary

Investigate
Teaching Suggestions and Sample Answers
Part A: Kinds of Weather Observations

1. Be sure that you read these steps with your students, so that they clearly understand their responsibilities in developing the weather center. Information sheets about the various weather instruments are provided in the back of this Teacher's Edition.

> **Assessment Tool**
>
> **Journal–Entry Evaluation Sheet**
> Use this sheet as a general guideline for assessing student journals, adapting it to your classroom if desired. You should give the **Journal–Entry Evaluation Sheet** to students early in the module, discuss it with them, and use it to provide clear and prompt feedback.
>
> **Journal–Entry Checklist**
> Use this checklist as a guide for quickly checking the quality and completeness of journal entries.

Investigation 1: Observing Weather

Research your weather observation.

a) Describe the science behind the weather observation in your journal.

b) What instrument, if any, is used to make your weather observation?

c) If you did not have a commercial version of your weather instrument, how could you make a homemade version? Draw a sketch of a homemade version in your journal.

3. You will be provided with information on how to make your weather observation properly so that your data are dependable.

Read the information carefully.

Inquiry
Writing a Protocol

A protocol is a procedure for a scientific investigation. It is a set of directions that someone else can read and follow. An important quality that your protocol should have is the ability to be replicated. In other words, your protocol should give consistent and reliable results, for anybody who uses it.

a) Write this information in the form of a protocol that others can easily understand and follow. Begin by writing a draft protocol for your weather observation. Remember to include the following in your protocol:

- the technique for taking the data;
- how to locate and set up your instrument properly, if your observation relies upon an instrument;
- how to read your instrument;
- how to record your data;
- the units of measurement to be recorded.

4. Exchange your draft protocol with one from another group in your class.

Read the other group's protocol.

a) Make comments on the other group's protocol. Your comments should be consistent with the criteria for a protocol outlined in Step 3 (a).

5. When you get your protocol back, revise it.

Practice making your weather observation using your protocol. Let each member of your group try this and compare the data you get from each person.

  Check your protocol with your teacher before proceeding to make observations.

a) Does data vary between group members? If yes, how can the variation be reduced?

Teacher Commentary

2. Assign the weather observation(s) that each group will make. Depending upon the number of groups in your class, you may find that a group can complete more than one measurement. For example, a group can measure both wind speed and direction, or observe both cloud type and cloud cover. You can assign additional observations if you wish, such as air pressure, humidity, etc.

   a) Provide students with the information sheets at the back of this Teacher's Edition (**Blackline Masters** *Climate and Weather* 1.1 – 1.7) to help them understand the science behind the weather observation. Provide students with other resources, such as books, CD-ROMs, or Internet access. Discuss with your students the science behind observing temperature, wind speed, wind direction, cloud types, cloud cover, precipitation type, precipitation amount.

   b) Thermometers are used to measure temperature, anemometers and wind vanes are used to measure wind speed and direction, respectively. Rain gauges are used to record the amount of precipitation, barometers measure air pressure, and hygrometers measure humidity. Other observations, like cloud cover or cloud type, can be made without instruments.

   c) In addition to the aforementioned sheets provided at the back of this Teacher's Edition, you can direct your students to the *Investigating Earth Systems* web site to view online instructions for constructing weather instruments.

3. Weather observation sheets (**Blackline Masters** *Climate and Weather* 1.1 – 1.7) provide information about making weather observations. Distribute as appropriate to your students.

   a) Student protocols should contain information that will help them make their weather observations accurately. Circulate around the room and check to make sure that students are providing correct information, because it is important that they understand how the weather instruments work and are used before they attempt to use them.

Teaching Tip

A protocol is a clear set of directions for how to do something. The protocol is written so that the results obtained by one person using it are just as dependable as the results obtained by another person following the same protocol. For example, at the doctor's office, pulse rate, body temperature, or blood pressure are all measured in the same way, by following a protocol for taking those measurements using specific tools and senses.

One of the first steps in most protocols that involve instruments is calibrating the instrument. This means that the instrument is first used to measure a value that is already known (the standard). For example, an instrument that measures the cloudiness of water can be calibrated by first using it to measure a sample of pure water. Since the water has 0 (no) cloudiness, the dials of the instrument can be set to a reading of 0 for pure water.

Another very important part of many protocols is sampling. Where, how, and when a sample is collected, as well as how it is stored before measurement, must be explained in a protocol.

> ### Teaching Tip
> If students are not familiar with the format of a protocol, you may find it useful to complete a sample protocol as a class. An example is shown below.
>
> **Sample Protocol for Weighing Yourself**
> **Time:** Same each morning, before breakfast
> **Site:** Hard, level surface, like a tiled bathroom floor
> **Conditions:** Wear the same garments each day for the weighing, so that the clothing-weight variable is controlled
> **Instrument:** Digital bathroom scale
> **Procedure:**
> 1. Follow the manufacturer's directions for zeroing the scale.
> 2. If the scale is being used for the first time, calibrate it by using it to weigh an object of known weight.
> 3. Adjust the scale, if necessary.
> 4. Carefully step onto the scale with your feet on either side of the readout window.
> 5. Read the weight (in pounds or kilograms) that the scale shows.
> 6. Get off the scale and store it in an area that is out of the way, where the scale will not be at risk from excess moisture or heat.
> 7. Record the time, weight, and date on a chart.

> ### Teaching Tip
> If any of your students have a parent or other adult in the healthcare profession, they may be able to get a sample of a protocol to share with the class. A scientist or technician would also be able to provide a sample protocol.

Monitor your students' progress throughout the course of **Investigation 1**. Look at students' journals. Provide positive feedback to students who record observations with clarity and detail. In a positive way, point out places where recorded observations might be improved.

4. Have students exchange their draft protocols and give them some time to review them.

    a) Students should make sure that the protocols contain all the information that is called for in **Step 3(a)** and that the information is easy to understand and follow. It is important to have each of the group members check the protocol to ensure reliability. The additional intergroup checking adds another layer of reliability to the protocol.

Teacher Commentary

5. On the basis of the comments made by their peers, students can make modifications to their protocols.

   a) Students should practice making their weather observations. Have them each make a few observations and then compare the data they get. It is likely that there will be greater variability between the first measurements, because students may not be familiar with the instrument they are using. They may find that they need to make their protocols more specific in order to reduce the variation in their data.

Teaching Tip

The accuracy of the instruments is measured by how close the instrument comes to the "correct" value. Governments set standards for different measures so that instruments can be calibrated to them. One standard that your students may be familiar with is Greenwich Mean Time. This is the 24-hour clock in Greenwich, England, by which all other clocks in the world are set.

Investigating Climate and Weather

INVESTIGATING CLIMATE AND WEATHER

6. Design a center to teach others about your weather observation.

   a) Make a sketch in your journal of how your center will look, and what questions your center will address. Some suggested questions are:

   - What does this weather observation tell you about the weather picture for a particular day?
   - What ideas do you have about how this weather measurement helps in predicting the weather?
   - In thinking about the protocol, what sorts of things could affect the accuracy of the weather data?

   Address any additional questions that you think will help others to understand your center.

7. Construct your center. Set it up where it will work best for visitors to make weather measurements or observations.

8. Work through the centers according to a schedule set by your teacher. Be sure that you are clear on how to follow each protocol. Also be sure that you understand what you are observing.

   a) Record your measurements in your journal.

   b) Write any questions that come to mind as you work through the center.

Part B: Making Observations for a Weather Map

1. Each student in the class will make weather observations at home for a week. The goal is to make a daily weather map of the local region.

   As a class, decide on a protocol for your observations. Here are some of the things you need to think about for the protocol:

   - An ideal weather map shows weather observations from many stations that are uniformly spaced.
   - Observations need to be made at exactly the same time of day.
   - Each observer needs to make exactly the same kinds of observations. Below is a list of these observations, with some helpful comments.

Teacher Commentary

6. Explain to students that the purpose of their center is to help others learn to make their weather measurement or observation accurately and dependably. You should tell your students where they will be setting up their centers, so they have an idea of the space available to them. Some centers will need to be near windows; others might need to be portable so that they can be carried outside.

Student centers should include:
- information on any weather instruments needed to make the measurement or observation
- the final protocol
- an explanation of the science behind the weather measurement or observation
- a data sheet for center visitors to record their observations
- questions for visitors to reflect upon, to help them understand the weather observation

7. You may want to require your approval of student centers before students begin to construct their centers.

Teaching Tip
You might find it useful to make centers out of cardboard boxes lying on their sides. One side can be cut out to leave the center open at the top. Protocols should be clearly written and labeled for the weather instrument. You can preserve the protocols by laminating them.

Wall posters paired with tables will also work for students to set up their stations. Posters may be better for your classroom situation than box centers, because they take up less room and are easier to store.

8. Students should cycle through each of the centers.
    a) Remind students to record all of the information they gain from each center in their journals.
    b) Students should record their questions in their journals.

Teaching Tip
It may take more than one class period for all groups of students to work through the centers. You may want to regulate the time spent by each group at each center.

> **Assessment Tool**
> Investigation Journal–Entry Evaluation Sheet
> Use this sheet to help students learn the basic expectations for journal entries that feature the write-up of investigations. It provides a variety of criteria that both you and your students can use to ensure that their work meets the highest possible standards and expectations. Adapt this sheet so that it is appropriate for your classroom, or modify the sheet to suit a particular investigation.

Part B: Making Observations for a Weather Map

1. Review the information provided in the student text and determine, as a class, how the weather observations will be made. Students need to make their observations for one week.

   a) Remind students to write the final protocol in their journals. You may want to type up the protocol and distribute it to the class.

   b) Students should record their data in a table like the one shown below.

Weather Element	Day 1	Day 2	Day 3	Day 4	Day 5
Wind					
Clouds					
Precipitation					
Temperature					
Pressure, etc.					

> **Teaching Tip**
> The data table shown here could be displayed on a large poster or on a class computer system. You could set it up on a whole-class server, if your computers have that capacity. You will also need to set up an observation schedule so that student groups know when they are responsible for taking measurements. Even with a schedule, you will need to remind students to take the measurements each day. It would also be a good idea for you to check the table on a daily basis, just in case one group's readings are extremely different from what the other groups have recorded.

Teacher Commentary

NOTES

Investigation 1: Observing Weather

Temperature: Your thermometer needs to be mounted in a shady place one to two meters above the ground. Make sure that it will not be blown away by a strong wind. Also make sure it is a shady spot. If you decide not to leave it outside the whole time, give it a few minutes to reach the outdoor temperature after taking it outside. Record temperatures in degrees Fahrenheit and degrees Celsius.

Wind direction: Record the wind direction as north, northeast, east, southeast, south, southwest, west, or northwest. Important: meteorologists record the direction the wind is blowing *from*, not the direction the wind is blowing to. For example, a northeast wind blows from the northeast toward the southwest.

Wind speed: You may not have an anemometer (an instrument that measures the speed of the wind), so you can use the Beaufort scale of wind speed. The Beaufort scale provides an estimate of wind speed based on observed effects of the wind. It is only approximate, but it is useful.

The Beaufort Wind Scale

Beaufort Number	Kilometers per hour	Miles per hour	Wind Name	Land Indication
0	<1	<1	calm	smoke rises vertically
1	1–5	1–3	light air	smoke drifts
2	6–11	4–7	light breeze	leaves rustle
3	12–19	8–12	gentle breeze	small twigs move
4	20–29	13–18	moderate breeze	small branches move
5	30–38	19–24	fresh breeze	small trees sway
6	39–50	25–31	strong breeze	large branches move
7	51–61	32–38	moderate gale	whole trees move
8	62–74	39–46	fresh gale	twigs break off trees
9	75–86	47–54	strong gale	branches break
10	87–101	55–63	whole gale	some trees uprooted
11	102–119	64–73	storm	widespread damage
12	>120	>74	hurricane	severe destruction

Teacher Commentary

NOTES

INVESTIGATING CLIMATE AND WEATHER

Clouds: Important types of clouds are pictured below. Observe which type or types of clouds are in the sky, and estimate how much of the sky is covered by clouds (zero-tenths, one-tenth, two-tenths, etc., up to complete cloud cover). If you are not sure about the types of clouds, write one or two sentences in your notebook to describe how they look.

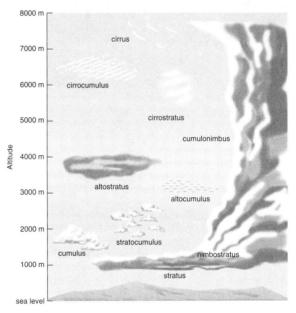

Precipitation: Is rain, drizzle, snow, or sleet falling? If so, is it light, moderate, or heavy? If it has rained since your last observation, how much rain has fallen? If it has snowed since your last observation, what is the depth of new snow? Record rainfall or snowfall in inches and centimeters.

- You need to decide upon a set of symbols for representing your weather observations on the map. The diagram on the following page shows the standard way of doing this on weather maps. Once you get used to this system, the data will show you at a glance what the weather was like at the station.

a) Once everyone has agreed on the protocol, write it down and make copies of it for each student.

b) Make your daily observations, and record them in your journal.

Teacher Commentary

Teaching Tip

The illustration on page C6 of the student text shows the location and shape of different types of clouds. Cloud types are reviewed in the **Digging Deeper** section of this investigation. You can find additional information and cloud images on the *Investigating Earth Systems* web site. Encourage students to write detailed descriptions and to draw sketches of the clouds they observe so that they will be able to use their data to specifically identify cloud type. You might point out to students that the illustration on page C6 is an artificial representation designed to show many kinds of clouds in a single illustration. It is unlikely that in the real world all of these kinds of clouds would exist together and be arranged as shown.

Investigating Climate and Weather

Investigation 1: Observing Weather

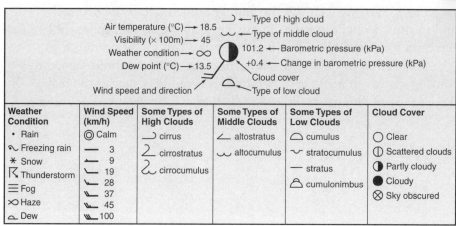

Weather Condition	Wind Speed (km/h)	Some Types of High Clouds	Some Types of Middle Clouds	Some Types of Low Clouds	Cloud Cover
• Rain	◎ Calm	⌒ cirrus	∠ altostratus	⌒ cumulus	○ Clear
Freezing rain	— 3	2 cirrostratus	⌄ altocumulus	⌄ stratocumulus	◐ Scattered clouds
* Snow	— 9	2 cirrocumulus		— stratus	◐ Partly cloudy
Thunderstorm	— 19			△ cumulonimbus	● Cloudy
≡ Fog	— 28				⊗ Sky obscured
∞ Haze	— 37				
⌒ Dew	— 45				
	— 100				

2. Your teacher will supply a base map for making each day's weather map. As a class, mark the locations on the map of all the "stations" that have been decided upon.

   After each day's observations, plot all of the data on the weather map for that day.

3. After all five daily weather maps have been plotted, have a class discussion that deals with the following questions. Record the results of your discussion in your journal.

   a) How much did the weather vary from place to place at the same time over the local region?

   b) Do you think that the variation seen on the maps reflects real variations in the weather, or were they caused by differences in the way observations were made? Explain.

   c) Give reasons for the changes you observed from day to day on your weather maps.

   d) Are there regions of the United States where typical day-to-day variations in the weather are greater than in your local region? Where they are less? What do you think might be the causes for these differences?

Inquiry
Making Observations

Every field of science depends upon observations. Meteorology starts with observations of the weather. It is important that all observations be made carefully and recorded clearly. In this investigation others in your class will be relying on the accuracy of your observations to draw a weather map.

Teacher Commentary

2. You may want to "blow up" a map of your town, county, city, or school district to poster size and have students stick a label containing their data directly on the map each day.

Teaching Tip

Additional information about the symbols used on weather maps can be found on the *Investigating Earth Systems* web site and in **Blackline Master** Climate and Weather 1.8.

3. Hold a class discussion in which students discuss the questions posed in the text. Answers to the questions will vary depending upon the region where you live and the time of year when you make your observations.

Teaching Tip

Observing weather is a longitudinal study. That means that the data must be collected over time. Each day, students will be adding weather data to the class chart. Remind them to look for relationships in the data. For example, when one thing changes, do others also change? Discuss any patterns that students find and what these might mean about weather. Encourage students to keep track of weather events all the way through the rest of the module.

Investigating Climate and Weather

INVESTIGATING CLIMATE AND WEATHER

As You Read...
Think about:
1. What does air temperature measure?
2. Why and how does wind blow?
3. How are clouds formed?
4. How is rain formed?

Digging Deeper

ELEMENTS OF WEATHER

All sciences begin with observations. Without observations, scientists have no way to develop new theories and to test existing theories. The weather is no exception. Meteorologists (scientists who study the weather) observe many elements of the weather. This takes place both at the Earth's surface and at high altitudes. The observations you made in this investigation include many of the most important elements of the weather. Weather observations are needed both for predicting the weather and for developing and testing new theories about how the weather works.

Air Temperature

Air consists of gas molecules, mostly nitrogen (N_2) and oxygen (O_2). Although you cannot see them with your eyes, the molecules are constantly moving this way and that way at very high speeds. As they move, they collide with one another and with solid and liquid surfaces. The temperature of air is a measure of the average motion of the molecules. The more energy of motion the molecules have, the higher the air temperature.

Air temperature is measured with thermometers. Common thermometers consist of a liquid-in-glass tube attached to a scale. The scale can be marked (graduated) in degrees Celsius or degrees Fahrenheit. The tube contains a liquid that is supplied from a reservoir, or "bulb," at the base of the thermometer. Sometimes the liquid is mercury and sometimes it is red-colored alcohol. As the liquid in the bulb is heated, the liquid expands and rises up in the tube. Conversely, as the liquid in the bulb is cooled, the liquid contracts and falls in the tube.

Evidence for Ideas

C 8
Investigating Earth Systems

Teacher Commentary

Digging Deeper

This section provides text and photographs that give students greater insight into different elements of weather and how scientists monitor these elements. You may wish to assign the **As You Read** questions as homework to help students focus on the major ideas in the text.

As You Read...

1. Air temperature is a measure of the average motion of gas molecules in the atmosphere. The greater the energy of motion of the molecules, the higher the temperature of the air.

   ### Teaching Tip
   Although the Kelvin temperature of an ideal gas is proportional to the energy of motion of the gas molecules, it is related only to the internal energy of motion. This energy of motion is calculated relative to the center of mass of the gas. The temperature of a gas container does not increase if we put the container on a moving train, nor does the temperature of the air increase because it is in motion as wind.

2. Wind blows because air pressure is higher in one place than in another place, and air moves from areas of higher pressure to areas of lower pressure. Also, the greater the difference in air pressure from one place to another, the stronger the wind.

3. Clouds are formed when humid air rises upward. As the air rises, it expands. As it expands, it becomes cooler. With enough cooling, water vapor in the air condenses into tiny water droplets, which are then visible as clouds.

4. Raindrops are formed when the cloud droplets grow large enough to fall out of the clouds.

   ### Assessment Opportunity
   You may wish to rephrase selected questions from the **As You Read** section into multiple choice or "true/false" format to use as a quiz. Use this quiz to assess student understanding and as a motivational tool to ensure that students complete the reading assignment and comprehend the main ideas.

About the Photo
The photograph of a standard inexpensive alcohol thermometer shows how the Celsius scale (on the left) and Fahrenheit scale (on the right) match at only one reading – negative 40 degrees. You might ask half of your students to convert negative 40 C to Fahrenheit, and the other half of your students to convert negative 40 Celsius to Fahrenheit. The results should match.

Investigating Climate and Weather – Investigation I

Investigation 1: Observing Weather

When you are measuring the air temperature, be sure to mount the thermometer in the shade. If the Sun shines on the thermometer, it heats the liquid, and the reading is higher than the true air temperature. Also, when you take the thermometer outside, give it enough time to adjust to the outdoor air temperature. That might take several minutes.

Wind

The wind blows because air pressure is higher in one place than in another place. The air moves from areas of higher pressure to areas of lower pressure. Also, the greater the difference in air pressure from one place to another, the stronger the wind. Objects like buildings, trees, and hills affect both the direction and speed of the wind near the surface. To get the best idea of the wind direction, try to stand far away from such objects. A park or a playing field is the best place to observe the wind.

Wind speed is measured with an anemometer. Most anemometers have horizontal shafts arranged like the spokes of a wheel. A cup is attached to the end of each shaft. The wind pushes the concave side of the cup more than the convex side, so the anemometer spins in the wind. The stronger the wind, the faster the cups spin. The cup spin rate is calibrated in terms of wind speed (e.g., miles or kilometers per hour).

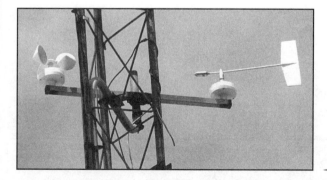

Teacher Commentary

About the Photo

The photograph shows a wind vane (on the right) and cup anemometer (on the left). Digital data sent through a cable to meteorologists are used to track changes in wind direction and wind speed.

INVESTIGATING CLIMATE AND WEATHER

You do not need an anemometer to estimate the wind speed. You can use a verbal scale, called the Beaufort scale (page C5). It describes the effect of the wind on everyday things like trees.

Wind direction is measured with a wind vane. You can also estimate the wind direction by yourself just by using your face as a "sensor." Face into the wind, and then record the direction you are facing.

Clouds

Clouds are formed when humid air rises upward. As the air rises, it expands and becomes colder. With enough cooling, water vapor condenses into tiny water droplets (or deposits into tiny ice crystals). The droplets or crystals are visible as clouds. Condensation is the change from water vapor to liquid water. Deposition is the change from water vapor directly to ice crystals. Condensation or deposition takes place when air is cooled to its dew-point temperature. When humid air is cooled at the ground (that is, when air reaches the dew point at ground level), fog is formed. You will learn more about clouds later in this module.

Clouds form at a wide range of altitudes, from near the ground to very high in the atmosphere. The appearance of clouds varies a lot, depending on the motions of the air as the clouds are formed. Other important things to observe about clouds are the percentage of the sky they cover, where they are located in the sky, how much of the sky they cover, and their direction of movement. A good way to find their direction of movement is to stand under a tree branch or an overhang of a building and watch the clouds move relative to that stationary object. Clouds move with the wind, so observing cloud motion provides information on the wind direction high in the atmosphere.

Teacher Commentary

About the Photo

Photograph of a roof-mounted wind vane. The arrow at the right side of the wind vane points into the wind.

Investigation 1: Observing Weather

Cirrus

Altostratus

Altocumulus

Nimbostratus

Cumulonimbus

Cumulus

Stratocumulus

Stratus

Teacher Commentary

About the Photo

There are four basic groups of clouds: high clouds, middle clouds, low clouds, and clouds with vertical growth. For descriptions of the various types of clouds, please refer to page 344 of this Teacher's Edition. The names of different clouds are based on the Latin word that describes their appearance from the ground: cirrus translates to "curl of hair," stratus means "layer," cumulus means "heap," and nimbus means "rain." Prefixes are added to cloud names for further description: "cirr" refers to high-level clouds and "alto" refers to middle-level clouds. Additional photographs of clouds are available through the *Investigating Earth Systems* web site.

INVESTIGATING CLIMATE AND WEATHER

Precipitation

Raindrops are formed when the cloud droplets grow large enough to fall out of the clouds. Most of the rain that falls in the winter, and even much of what falls in the summer, is from melting of snowflakes as they fall through warmer air.

Rainfall is measured by the depth of water that falls on a level surface without soaking into the ground. Rainfall is measured with a rain gauge. A basic rain gauge is nothing more than a cylindrical container, like a metal can with a flat bottom, that is open to the sky. The only problem is to get an accurate measurement of the depth of water that has fallen. Accurate rain gauges are designed so that the water that falls into the container is funneled into a much narrower cylinder inside. In that way, the depth of the water is magnified, and is easier to read.

If you live in a part of the United States where it snows in winter, you can easily measure the snow depth with a ruler graduated in either centimeters or inches. The best time to make the measurement is right after the snow stops falling. The measurement can be tricky, because wind can cause snow to drift. The best place to measure snow depth is on level ground far away from buildings and trees. Take measurements at several spots and compute an average.

Teacher Commentary

NOTES

Investigation 1: Observing Weather

Review and Reflect

Review

1. List the weather instruments you studied in this investigation and describe what each measures.
2. What factors do you need to consider to get an accurate reading of each of the following:
   a) air temperature?
   b) wind speed and direction?
   c) rainfall? snowfall?

Reflect

3. What are some drawbacks to relying on weather measurements taken by your class, as opposed to following the reports issued by professionals?
4. Most of the weather observations you made can be made and recorded automatically by instruments. Using one kind of observation as an example, describe an advantage and disadvantage of using technology to record weather information versus people recording the information.

Thinking about the Earth System

5. How do seasonal changes in air temperature affect plants and animals?
6. How does the wind interact with the surface waters of the ocean or large lake?
7. How does the wind affect trees, bushes, and other vegetation?
8. How does air temperature and wind speed influence how comfortable you are when outside?

Thinking about Scientific Inquiry

9. How is writing and following a protocol important to the inquiry process?
10. How did you use mathematics in this investigation?
11. Describe an example of how you collected and reviewed data using tools.

Teacher Commentary

Review and Reflect

Review

Now that your students have had a chance to become specialists at one kind of weather observation, and have also had the opportunity to compile data from throughout their community, they have begun to think about what is required to produce a weather report.

The answers provided below are for you, the teacher. It is not expected that your students will answer with the same level of sophistication. Use your knowledge of the students as well as the standards set by your school district to decide what answers you will accept. In student answers, look for evidence of an understanding of the processes involved, as well as for any misconceptions that still remain. Encourage students to express their ideas clearly and to use correct science terminology where appropriate.

1. Answers will vary, depending upon the observations that you have your students make. Most likely, they will use:
   - thermometers to measure temperature
   - anemometers to measure wind speed
   - wind vanes to measure wind direction
   - rain gauges to measure precipitation
   - barometers to measure air pressure
   - hygrometers to measure humidity
   - their own senses to make observations of cloud cover and types

2. a) When measuring air temperature, students should make sure that the thermometer is mounted well above the ground in an area where it is not exposed to direct sunlight. Exposure to sunlight could cause a thermometer reading that is unrepresentative of the true air temperature.

   b) To measure wind direction, students should know which direction is north, so that the wind vane can be oriented properly. To measure wind speed, an accurately calibrated anemometer of some kind is needed. An alternative is to use the Beaufort wind scale to make an approximate measurement of wind speed.

   c) To get an accurate reading of precipitation that falls as rain, all that is needed is a suitably calibrated rain gauge, placed well away from buildings and trees. Measuring snow depth is more difficult, because the snow is likely to blow into drifts if the wind is strong during or after the snowfall. A site on open, level ground far away from buildings and trees is best.

Reflect

Give your students adequate time to review and reflect on what they have done and understood in **Investigation 1**. Ensure that all students think about and discuss the questions listed here. Be on the lookout for any misunderstandings; where necessary, help students clarify their ideas.

3. Students will likely respond that they are not professionally trained to make weather observations, the way meteorologists are. Also, meteorologists have access to much more sophisticated and accurate instruments for making weather observations.

4. Answers will vary depending upon which weather observation students select

Thinking about the Earth System

It is very important that students begin to relate what they are studying to the wider idea of the Earth System. This is a complex and largely inferred set of concepts that students cannot easily understand from direct observation. Remember, the goal is that students will have a working understanding of the Earth System by the time they complete eighth grade. Although it can be taught as a piece of information, true understanding is largely dependent upon comprehending how numerous specific Earth-science concepts connect with the idea of the Earth as a system. Be sure to spend some time helping students to make what connections they can between the focus of their investigations and this wider aspect.

5. In temperate regions, including most of the United States (outside of Hawaii, southern Florida, and northern Alaska), annual plants make their seeds in the warm season and die at the beginning of the cold season, and perennial plants grow in the warm season and become dormant in the cold season. Deciduous trees and shrubs lose their leaves in the cold season. Some animals accommodate to the seasons by hibernating or migrating south in winter and north in summer; others are adapted to year-round life in a particular region.

6. Wind generates waves on the surface of lakes or the ocean. The stronger the wind, the larger the waves. Wind also generates surface currents.

7. Wind can have a negative effect on plants by blowing trees over and pulling the roots of vegetation from the ground. High wind can also affect the growth and shape of trees and other plants. Wind can also positively affect plants, by dispersing seeds and pollen.

8. Obviously, the higher the temperature, the warmer you feel. Humidity is an important factor as well, especially at high temperatures, because high humidity decreases the rate at which perspiration evaporates, making your cooling system less efficient. The wind has two effects: it increases the rate at which perspiration evaporates; and when the air temperature is less than your body temperature, it increases the rate of heat loss from your body.

Teacher Commentary

Thinking about Scientific Inquiry

The many processes and skills involved in scientific inquiry can be taught as pieces of information, but for a solid understanding, students need considerable firsthand investigative experience. Students are given many opportunities to think about the connections between their investigations and inquiry processes.

9. Writing and following a protocol is important to the inquiry process because it guides the process of planning an investigation, and it improves the quality and consistency of data by describing how to conduct the investigation and how to collect and review the data.

10. Answers will vary depending upon the measurements made by the students.

11. Answers will vary depending upon the example students choose to describe.

Assessment Tool

Review and Reflect Journal-Entry Evaluation Sheet
Depending upon whether you have students complete the work individually or within a group, use the **Review and Reflect** part of the investigation to assess individual or collective understandings about the concepts and inquiry processes explored. Whatever choice you make, this evaluation sheet provides you with a few general criteria for assessing content and thoroughness of student work. Adapt and modify the sheet to meet your needs. Consider involving students in selecting and modifying the assessment criteria.

Teacher Review

Use this section to reflect on and review the investigation. Keep in mind that your notes here are likely to be especially helpful when you teach this investigation again. Questions listed here are examples only.

Student Achievement

What evidence do you have that all students have met the science content objectives?
_____

Are there any students who need more help in reaching these objectives? If so, how can you provide this? _____
_____

What evidence do you have that all students have demonstrated their understanding of the inquiry processes? _____
_____

Which of these inquiry objectives do your students need to improve upon in future investigations? _____
_____

What evidence do the journal entries contain about what your students learned from this investigation? _____
_____

Planning

How well did this investigation fit into your class time? _____
_____

What changes can you make to improve your planning next time? _____
_____

Guiding and Facilitating Learning

How well did you focus and support inquiry while interacting with students?
_____

What changes can you make to improve classroom management for the next investigation or the next time you teach this investigation? _____
_____

Teacher Commentary

How successful were you in encouraging all students to participate fully in science learning? _____

_____

How did you encourage and model the skills values, and attitudes of scientific inquiry? _____

_____

How did you nurture collaboration among students? _____

_____

Materials and Resources

What challenges did you encounter obtaining or using materials and/or resources needed for the activity? _____

_____

What changes can you make to better obtain and better manage materials and resources next time? _____

_____

Student Evaluation

Describe how you evaluated student progress. What worked well? What needs to be improved? _____

_____

How will you adapt your evaluation methods for next time? _____

_____

Describe how you guided students in self-assessment. _____

_____

Self Evaluation

How would you rate your teaching of this investigation? _____

_____

What advice would you give to a colleague who is planning to teach this investigation? _____

_____

NOTES

Teacher Commentary

INVESTIGATION 2: COMPARING WEATHER REPORTS

Background Information

Weather Forecasting

There are many approaches to forecasting (predicting) the weather. Some are very simple; some are much more sophisticated and require powerful computers. None of the methods is anywhere near perfect. Weather forecasting will always involve some degree of uncertainty. Even if all the processes involved in the weather were completely understood, and could be built into computer programs for weather forecasting, there would still be two major problems: (1) The grid spacing of weather observations would still be larger than important local weather systems like thunderstorms and tornadoes. (2) Like many physical systems, the weather is chaotic, in the sense that very small irregularities in the system at a given time tend to develop into much larger irregularities in ways that are inherently impossible to predict.

The simplest way of predicting the weather is to assume that future weather will be the same as present weather. That effect is called persistence of weather. For periods up to several hours, such predictions have good accuracy.

The weather can also be predicted by what might be called the steady-state method. The assumption is made that weather systems tend to move in the same direction and at the same speed without changing. That method also works well for short time periods of up to several hours.

Another method of forecasting, called the analog method, makes use of the fact that the pattern of weather systems on a weather map might be much like a kind of pattern that produced a given weather condition at various times in the past. In that way, earlier weather patterns can be used to make predictions of future weather patterns. That method is far from perfect, however, because no two weather patterns are exactly the same, and as the different patterns evolve, they typically become less and less similar.

In recent years, computer models of the weather have come into common use in predicting the weather. The computer model makes use of a number of equations (mostly partial differential equations, in the parlance of applied mathematics) that are supposed to describe the physics of the atmosphere. The starting conditions of the atmosphere (based on the latest set of weather observations from stations throughout the Northern Hemisphere) are fed into the computer, and the program lets the atmospheric system run in order to predict how the state of the atmosphere will change over some small time step. That new set of conditions is then taken as the starting condition for the next time step, and so on, for a very large number of time steps adding up to many hours or many days. The weather forecaster uses the results, along with his or her experience and judgment, to make a forecast. There are two problems here: (1) Again, the grid spacing of the observations is never ideally close. (2) It is still difficult to build all of the various physical processes of the atmosphere into a neat set of equations to solve; the physics of clouds and precipitation are particularly difficult to deal with.

More Information…on the Web
Go to the *Investigating Earth Systems* web site www.agiweb.org/ies for links to a variety of web sites that will help you deepen your understanding of content and prepare you to teach this investigation.

Teacher Commentary

Investigation Overview

Students compare weather reports from a variety of sources, evaluating them for the information they contain as well as for their accuracy. They work as a class to make a table of weather terms that can serve as a reference throughout the rest of the module. **Digging Deeper** reviews what information is commonly found in a weather report and in a weather forecast. The reading also reviews the history of weather forecasting, and explains why weather prediction will never be without error.

Goals and Objectives

As a result of **Investigation 2**, students will understand what information is contained in weather reports and forecasts and which types of weather reports or forecasts are most appropriate in different situations.

Science Content Objectives

Students will collect evidence that:
1. Certain weather words are used to describe the quality of weather measurements.
2. Weather reports are descriptions of weather conditions.
3. Weather forecasts are predictions of the weather, over both the short term and the longer term.
4. Weather reports vary, depending upon their sources of information, audience, and method of transmittal.
5. Greater understanding of the science of weather, along with improvements and innovations in technology, have allowed increasingly accurate weather forecasts. However, weather prediction will never be an exact science.

Inquiry Process Skills

Students will:
1. Predict the accuracy of different weather reports.
2. Compare weather reports from different sources.
3. Record observations.
4. Evaluate weather reports for level of information and accuracy.
5. Recognize patterns and relationships in weather reports.
6. Use data to support or refute predictions.
7. Conduct research on weather terms.
8. Arrive at conclusions.
9. Communicate observations and findings to others.

Connections to Standards and Benchmarks

In **Investigation 2**, students compare weather reports from a variety of sources. This will help them begin to think about the information contained in different weather reports and to consider which reports are most accurate. These observations contribute to developing the National Science Education Standards and AAAS Benchmark shown below.

NSES Links
- Different kinds of questions suggest different kinds of scientific investigations. Some investigations involve observing and describing objects, organisms, or events; some involve collecting specimens; some involve experiments; some involve seeking more information; some involve discovery of new objects and phenomena; and some involve making models.

- The atmosphere is a mixture of nitrogen, oxygen, and trace gases that include water vapor. The atmosphere has different characteristics at different altitudes.

AAAS Link
- Because the Earth turns daily on an axis that is tilted relative to the plane of the Earth's yearly orbit around the Sun, sunlight falls more intensely on different parts of the Earth during the year. The difference in heating of the Earth's surface produces the planet's seasons and weather patterns.

Teacher Commentary

Preparation and Materials Needed

Preparation

In **Investigation 2**, your students will examine the differences in content and format of various weather reports. Some content is determined by the delivery system of the report. Television and Internet reports can show moving images, but newspaper reports are limited to static images. The most limited reports are those on the telephone or radio, because they are restricted to the spoken word.

You will need to collect weather reports and forecasts from a variety of sources for students to analyze. If possible, and if your school allows, you can record television reports on video. If this is not possible, you may want to have a television in your room and have students access the Weather Channel directly.

You will also need to collect weather forecasts and weather data for each of three consecutive days. Students will compare the actual data with the forecasts to determine the accuracy of the weather forecasts.

Materials

- sample weather report/forecast
- weather reports/forecasts for three consecutive days, from the following media: television, national newspaper, local newspaper, commercial or public radio, the NOAA weather radio, Internet, telephone*
- weather data for the same three consecutive days as used for the weather forecasts (temperature, wind speed, precipitation, cloud cover, humidity, etc.)
- reference resources about weather

---

* The *Investigation Earth Systems* web site provides suggestions for obtaining these resources.

Investigating Climate and Weather

INVESTIGATING CLIMATE AND WEATHER

Investigation 2:

Comparing Weather Reports

Key Question
Before you begin, first think about this key question.

What does a weather report contain?

Think about what you already know about weather reports. What information is contained in one? What do you need to know to interpret a weather report?

Share your thinking with others in your group and with your class.

Materials Needed

For this investigation, your group will need:

- sample weather report
- weather reports or forecasts for three consecutive days, from one of the following media: television, national newspaper, local newspaper, commercial or public radio, the NOAA weather radio, Internet, telephone
- weather data for each of the three days following the reports (temperature, wind speed, precipitation, cloud cover, humidity, etc.)
- reference resources about weather

Investigate
1. A *weather report* tells you what the weather conditions are at the time. A *weather forecast* tells you what the weather is likely to be in the future. Your group will receive one of several different kinds of weather reports.

   a) Make a list of all the weather words in the report.

C 14
Investigating Earth Systems

Teacher Commentary

Key Question

Write the **Key Question** on the blackboard or on an overhead transparency. Encourage students to think about what they learned in **Investigation 1** regarding the methods used to observe weather, and to connect what they learned to the information contained in weather reports. Tell students to record their ideas in a new journal entry.

Discuss students' ideas. Ask for a volunteer to record responses on the blackboard or overhead projector so that you can circulate among the students, encouraging them to copy the notes in an organized way.

Student Conceptions about Information in Weather Reports

Students are likely to have some knowledge of weather reports from television and radio. They, like everyone else, are often eager to know what the weather is going to be like so that they can plan accordingly. However, they may focus only on certain aspects of weather, like temperature and precipitation. If they live in a region that experiences high humidity, they may know how this is reported. Students may be less certain about wind direction and speed, and they may have very little understanding of dewpoint, barometric pressure, fronts, and high and low pressure systems.

Assessment Tool

IES **Key–Question Evaluation Sheet**
Use this evaluation sheet to help students to understand and internalize basic expectations for the warm-up activity.

About the Photo

The photo depicts a television weatherperson in front of a satellite image of a hurricane. Such images are often shown in animation, in a time-lapse mode that shows the movement and change in weather systems and weather patterns over periods of several hours in just seconds of broadcast time. Such images are very expressive and instructive of the true behavior of weather systems.

Investigation 2: Comparing Weather Reports

b) Turn the list into a table like the one below. If someone in your group knows some of the other information for the table, fill in that part. If not, leave it blank for later. If you find a term that you do not understand, write it down as a question to answer later.

Weather Word	Descriptors Used	Definition	Instrument Used for Measuring	Unit of Measurement
Wind Speed				
Wind Direction				
Clouds				
Kind of Precipitation				
Amount of Precipitation				
Temperature				
Pressure				
Other				

2. Study your group's report a second time.

   a) This time, make a table that shows what kind of weather information your report includes.

   The table should clearly show the kinds of weather reports your class is studying (newspaper, radio, and so on),

Teacher Commentary

Investigate

Teaching Suggestions and Sample Answers

From their experiences in **Investigation 1**, most students should be getting used to working in groups. Be alert to any problems in this regard, re-forming groups if necessary.

1. Depending upon your resources, you might have a full set of weather reports for each group of students, or you may have master sets that can circulate from group to group. Review with your students the difference between a weather report and a weather forecast.

   a) Students should list as many weather words as are shown in the report or forecast. They can use the table in the text as a guide, and to give them ideas about what to look for.

   b) Circulate among your students, noting the kind of information they are recording about the weather reports. This can give you an informal pre-assessment of their understanding of weather-related vocabulary terms.

   Teaching Tip

   Students can use the table on page C15 of the student text as a guide to get them started. An example of this table is provided as a **Blackline Master** (*Climate and Weather* 2.1) at the back of this Teacher's Edition. Their tables should be similar to the one shown in the text; however, they will need to make the data columns wider, particularly the column containing definitions for the words.

2. Students should now make a second data table that looks at the kind of information found in their weather report. They can use the table on page C16 of the student text as a guide. A copy of this table is also provided as a **Blackline Master** (*Climate and Weather* 2.1) at the back of this Teacher's Edition.

   a) The information found on the weather reports will vary. Students may note:
   - temperature
   - cloud cover
   - air pressure
   - precipitation
   - fronts
   - high and low air pressure systems
   - dewpoint
   - wind-chill factor
   - relative humidity
   - wind speed and direction

   b) Students should share their findings and compile a complete list. You may wish to make the table into an overhead transparency and fill it in as a class.

Teaching Tip
You may wish to have your students make a table on the computer. One important science inquiry skill you can remind your students about is the collecting and reviewing of data using tools. Both the table and the computer are data management tools.

Teacher Commentary

NOTES

INVESTIGATING CLIMATE AND WEATHER

the categories of weather information the report includes (temperature, wind speed, etc.), and the date and time the report was produced.

b) Share your findings with other groups, in order to fill in your table completely.

Weather Report Source	Date & Time of Report	Temperature	Wind Speed	Humidity	Cloud Cover	Other
Local Paper						
National Newspaper						
Radio						
Internet						
Television						
Phone						

Inquiry

Collect and Review Data Using Tools

One important science inquiry skill is the collecting and reviewing of data using tools. In this inquiry you are using tables as data management tools. You could also use a computer as a tool to make your table.

3. Obtain the weather forecasts (temperature, wind speed, precipitation, etc.) from your weather source during a three-day period.

   a) Record the forecasts in the form of a table.

   b) Record also the actual weather data.

   c) Share your findings with other groups, in order to fill in your table completely.

The following is an example of a table that you can use:

Source	Day 1 Forecast	Day 1 Actual Data	Day 2 Forecast	Day 2 Actual Data	Day 3 Forecast	Day 3 Actual Data
Local Paper						
National Newspaper						
Radio						
Internet						
Television						
Phone						

Teacher Commentary

Assessment Tools

Journal–Entry Evaluation Sheet
Use this sheet as a general guideline for assessing student journals, adapting it to your classroom if desired.

Journal–Entry Checklist
Use this checklist as a guide for quickly checking the quality and completeness of journal entries.

3. Students should now look at weather forecasts for three consecutive days. You should also provide them with actual weather data that corresponds to the days for which the weather was forecast. You can either provide students with a copy of each kind of weather forecast or have each group work on weather forecasts from one source. Students can compile the data as they did in **Step 2**.

   a) Students can use the table at the bottom of page C16 of the student text as a guide. An example of this table is provided as a **Blackline Master** (*Climate and Weather* 2.3) at the back of this Teacher's Edition.

   c) Students should share their findings and compile a complete list. You may wish to make the table into an overhead transparency and fill it in as a class.

   d) Students should attempt to determine which weather-forecast source is most accurate. The source they think is most accurate should be ranked "1." Suggest to students that they add another column to their table labeled "ranking."

   e) Students should explain why they ranked the sources in the order that they did. The best evidence to support their rankings is to compare the forecast data with the actual weather data and rank the best match as "1" (for most accurate).

Teaching Tip

This section of **Investigation 2** helps your students understand what is meant by the term accuracy. Weather reports and forecasts afford a special opportunity to compare predictions with actual data. Comparing weather forecasts with weather data is one way of establishing the concept of accuracy.

> ### Making Connections...*with mathematics and technology*
> Students can construct and complete their data tables using a computer. The ranking that students are asked to do in **Step 3d)** is a direct mathematics connection. You may want to work collaboratively with your students' mathematics teachers to arrange for the concept of ranking to be reinforced in math class.

Investigation 2: Comparing Weather Reports

d) Compare the predicted weather with the actual weather, to judge the accuracy of each of the weather forecasts. Look over the information in your table carefully. Rank the weather-forecast sources from most accurate to least accurate. You can do this by giving the number 1 to the most accurate report, 2 to the second, and so on.

e) What evidence do you have to support your rankings?

4. Discuss the following questions within your group. Record the results of your discussion in your journal. (Report in these questions means report or forecast.)

   a) Which kind of weather report has the most information?
   b) How does the format of the report make that possible?
   c) Which kind of weather report has the least information? Explain.
   d) What information do all of the weather reports include?
   e) Why do you think that this information is included in every weather report?
   f) Which kind of weather report best helps you to understand patterns in the weather? Why?
   g) Which weather report would be most helpful for planning an outdoor event three days in the future? Why?
   h) Which weather report would be most helpful if you were traveling to another part of the country? Why?

5. After discussing the weather report questions, share your ideas with the whole class.

   a) What kinds of information need to be included in an accurate and useful weather report?
   b) How do you think you would obtain the information that you need to make a more complete and accurate weather report?

6. Go back to your table of weather terms from the beginning of the investigation.

   Divide up the terms in the table equally among your group members. Use the resources in your classroom, library, and at home to complete the information in the table.

   a) Complete the table in your journal.
   b) Write down any questions you think of as you do your research.

Inquiry

Predictions in Science

Scientists make predictions and justify these with reasons. A meteorologist uses all the data available to make a complete picture of the present and future weather. By using these data, and his or her knowledge of how weather systems form and move, the meteorologist can predict or forecast the weather.

Teacher Commentary

4. Answers to these questions will vary depending upon the sources of the weather reports you provide to your students. Students should base their answers on the data tables they have produced during this investigation, and you should encourage them to support their responses using their findings.

5. a) Answers will vary, but most weather reports minimally include information about high and low temperatures, and precipitation. See **Digging Deeper**.

   b) Answers will vary. One possible response is that one could contact an organization like the national weather service to gather additional information not included in a basic weather report. Examples of additional information include regional atmospheric pressure, wind speed, etc.

6. Provide students with reference resources about weather to help them to define their weather terms. If your students have computer access, the *Investigating Earth Systems* web site can serve as a valuable resource.

A sample, inexhaustive table is provided below.

Weather Word	Descriptors Used	Instrument Used for Measuring	Unit of measurement
Wind speed	Miles per hour, breezy, calm, etc.	Anemometer	Miles per hour, kilometers per hour, meters per second
Wind direction	Northeast wind, from the west, etc.	Wind vane	Compass directions (N, NE, E, etc.)
Clouds	Mostly cloudy, partly sunny, clear, etc.	Direct human observations, radar, satellites	Primarily qualitative, but cloud cover is sometimes reported in percentage of cloud cover
Kind of precipitation	Rain, hail, sleet, etc.	Direct human observations, radar, satellites	
Amount of precipitation	Inches of rain, inches of snow, etc.	Rain gauge, snow gauge	Inches, feet (snow only), sometimes millimeters
Temperature	Degrees, very cold, hot, etc.	Thermometer	Degrees F, degrees C
Pressure	High pressure, low pressure, inches of mercury	Barometer	Inches of mercury, millibars
Other			

Teaching Tip

Have the class compile their final table of weather words on a large piece of poster board. Post the table in the classroom for quick and easy reference throughout the rest of the module.

INVESTIGATING CLIMATE AND WEATHER

c) Make a class table of information about weather words by having each group contribute the words it has found. Keep this table posted as a reference for the rest of the investigations in this module. Keep a copy in your journal

As you work through the other investigations in this module, you will probably find other words related to weather and climate. Add these to the table as you accumulate them.

As You Read...
Think about:
1. What is the difference between a weather report and a weather forecast?
2. What information is contained in a detailed weather report?
3. What does normal average temperature mean?
4. How has weather forecasting changed during the past two hundred years?

Digging Deeper

WEATHER REPORTS AND FORECASTS
Weather Reports

Different weather reports contain different amounts of information. The simplest and shortest weather report contains only one piece of information, the present temperature. This kind of report you often hear on the radio. More detailed weather reports contain information about precipitation, wind speed and direction, relative humidity, atmospheric pressure, and so on.

1 TODAY MONDAY, JUNE 11

Chicago: A good deal of sunshine, breezy and warm. Humidity continues slowly upward, and a passing late-afternoon or evening thunderstorm cannot be ruled out. Partly cloudy and mild overnight.

HIGH | LOW
84 | 64

A typical weather report tells you the highest and lowest temperatures for the past day. The day's lowest temperature usually occurs just after sunrise. The day's highest temperature is usually reached during early to mid afternoon. A weather report also tells you the present temperature. It may also give you the average temperature for the day. The average daily temperature lies halfway between the highest temperature and the lowest temperature. The weather report might also tell you how

Teacher Commentary

Digging Deeper

This section provides text, an illustration, and a photograph that give students greater insight into the topic of weather reports and forecasts. You may wish to assign the **As You Read** questions as homework to help students focus on the major ideas in the text.

As You Read...

1. A weather report gives information about past or current weather conditions; a weather forecast (short term) gives predictions of what the weather will be like over the course of a few days.

2. A detailed weather report contains information about precipitation, wind speed and direction, relative humidity, atmospheric pressure, etc.

3. The normal average temperature of any given day is found by taking the average of all the individual average temperatures for that calendar day over the past 30 years.

4. Greater understanding of the science of weather, along with improvements and innovations in technology, have allowed increasingly accurate weather forecasts. However, weather prediction will never be an exact science.

Assessment Opportunity

You may wish to rephrase selected questions from the **As You Read** section into multiple choice or "true/false" format to use as a quiz. Use this quiz to assess student understanding and as a motivational tool to ensure that students complete the reading assignment and comprehend the main ideas.

Investigation 2: Comparing Weather Reports

many degrees the average temperature is above or below the normal temperature for that day. The normal temperature is found by averaging the average temperatures for the calendar day for the past 30 years.

Most weather reports give the amount of precipitation (rain or melted snow), if any, that fell during the past day. They also tell you the totals for the current month and the current year. Reports also indicate how much the monthly and annual precipitation totals are above or below normal (the long-term average).

Weather Forecasts

Most people are interested in what the weather will be tomorrow or in the next few days. Predictions of the weather for up to a week in the future are called short-term forecasts. Meteorologists also try to make long-term forecasts (called "outlooks") of the weather for a month, a season, or a whole year. Long-range outlooks are different from short-term forecasts in that they specify expected departures of temperature and precipitation from long-term averages (e.g., colder or warmer than normal, wetter or drier than normal).

In earlier times, before invention of the telegraph and the telephone, weather observations from faraway places could not be collected in one place soon after they were made. In those times, the only way of predicting the weather was to use your local experience. Given the weather on a particular day, what kind of weather usually follows during the next day or two? As you can imagine, the success of such forecasting was not much better than making a random guess.

Beginning in the 1870s, a national weather service used the telegraph to gather weather observations from weather stations located over large areas of the country. Simultaneous weather observations allowed

Teacher Commentary

NOTES

INVESTIGATING CLIMATE AND WEATHER

meteorologists to plot weather maps and follow weather systems as they moved from place to place, greatly improving the accuracy of weather forecasts.

Through the 20th century, meteorologists developed even better tools for observing and predicting the weather. As you will learn in the following investigations, special instruments measure weather in the atmosphere far above the ground. Satellites orbiting the Earth send back images of the weather over broad areas of the planet. In addition, computer models were developed for weather forecasting. The important processes operating in the Earth system that govern weather are built into a computer model. The model starts with the present weather and tries to simulate how the weather will evolve in the future. These computer models are run on supercomputers. They can handle enormous amounts of observational data and make billions of computations quickly. Today's computer models do a very good job of predicting the weather for the next few days. You know, however, that sometimes the forecast is wrong! The science of weather forecasting is still developing.

Weather prediction will never be perfect. One reason is the absence of reliable weather observations from large areas of the globe (especially the oceans). These observations are needed for computer models to accurately represent the present state of the atmosphere. A second reason is that even small changes in the weather in one place can cause much larger changes in weather elsewhere. The effects are small at first, but they become much greater. It's very difficult for computers to simulate these interactions. Although forecasts will never be perfect, they will continue to improve in the years ahead. Through research, meteorologists learn more and more about the details of how weather in the Earth system works.

Teacher Commentary

About the Photo

The photo on page C20 of the student text shows high cirrus and cirrostratus clouds at sunset. Sunsets and sunrises are sometimes vividly yellow, orange, and red because the atmosphere filters out the blue component of sunlight more strongly than the red component. When the Sun is low in the sky, the sunlight has to travel a much greater distance through the atmosphere, causing this differential filtering effect to be much greater.

Investigation 2: Comparing Weather Reports

Review and Reflect

Review

1. What was the most accurate weather forecast that your class studied? Explain.
2. Where do the weather reports and forecasts that your class studied come from?
3. Why does the accuracy of weather forecasts usually decrease as the number of days ahead increases?

Reflect

4. If you were to write your own weather report, what would you include and how would you get your information?
5. Why do you think weather reports vary over a given area?
6. Which kind of weather report is most useful to you on a daily basis? Why is that?

Thinking about the Earth System

7. How has communication technology helped to develop a better understanding of the Earth system? Give an example.
8. Write down all the connections you can think of in this investigation that show a relationship between weather and the Earth's systems. Keep this record on your *Earth System Connection* sheet.

Thinking about Scientific Inquiry

9. What question did you explore in this investigation? Can you answer the question now? Explain.
10. What tools did you use to collect and review data in this investigation?

Teacher Commentary

Review and Reflect

Review

Before completing **Investigation 2**, many students may have had few ideas about weather reporting and forecasting. Help your students see that although science investigations provide some answers to scientific questions, they often raise further questions for investigation. Spend some time having students talk about these possible questions.

1. This answer will depend upon the weather forecasts that your class looked at.

2. Answers will vary, depending upon the weather forecasts that your class looked at.

3. Because weather prediction is not perfect, the accuracy of weather forecasts usually decreases as the time period of the forecast (i.e., the number of days ahead) increases. Small changes in the weather in one place can cause much greater changes in weather elsewhere. It is very difficult for computers to simulate how changes in one location will affect weather in another location.

Reflect

Give students adequate time to review and reflect on what they have done and understood in **Investigation 2**. Ensure that all students think about and discuss the questions listed here. Be alert to any misunderstandings. Where necessary, help students clarify their ideas.

4. Answers will vary, but will likely include information about temperature, precipitation, wind speed and direction, relative humidity, atmospheric pressure, etc.

5. Most weather reports will contain information about temperature, but different areas may have different opinions on what aspects of the weather are most important to them. For this reason, weather reports sometimes vary over a given region. For example, in areas near the coast where fishing and boating are important activities, the weather report might include a marine forecast that gives information on the tides, wave heights, and wind. A short distance inland in a farming region, the tides are not important to the people using the forecast, so this information might be left out so that other information can be discussed in greater detail.

6. Answers will vary, but students are likely to say that a weather report that gives current temperature and precipitation is most useful.

Thinking about the Earth System

Give students time to make any connections they can between what they have done in **Investigation 2** and the Earth System. Refer students to the diagram on page Cviii of the student text as a means of exploring the Earth System further. Ensure that students add their ideas to their *Earth System Connection* sheets.

7. There are many possible answers. One example is that instantaneous transmission of large volumes of weather data, including satellite images and radar images, helps to show the interaction between the atmosphere and the hydrosphere.

8. Students should be able to list several Earth system connections. One example would be the effect that weather (atmosphere) has on humans and other animals (biosphere).

Thinking about Scientific Inquiry

In **Investigation 2**, your students have been exploring questions about weather reporting and forecasting. Your students may need help in seeing that scientists make predictions and justify their predictions with reasons, using all the data available. Have your students refer to the fourth **Inquiry Process** listed on page Cx of the student text: "Collect and review data using tools." Ask them to look carefully at the explanation given for this inquiry process.

9. Answers to this question will vary.

10. Students used tables, weather reports, weather forecasts, and weather data. They may have also used computers.

Assessment Tool

Review and Reflect Journal–Entry Evaluation Sheet
Use the general criteria on this evaluation sheet for assessing content and thoroughness of student work. Adapt and modify the sheet to meet your needs. Consider involving students in selecting and modifying the criteria for evaluating their reflections on **Investigation 2**.

Teacher Commentary

NOTES

Teacher Review

Use this section to reflect on and review the investigation. Keep in mind that your notes here are likely to be especially helpful when you teach this investigation again. Questions listed here are examples only.

Student Achievement

What evidence do you have that all students have met the science content objectives?

_____

Are there any students who need more help in reaching these objectives? If so, how can you provide this? _____

_____

What evidence do you have that all students have demonstrated their understanding of the inquiry processes? _____

_____

Which of these inquiry objectives do your students need to improve upon in future investigations? _____

_____

What evidence do the journal entries contain about what your students learned from this investigation? _____

_____

Planning

How well did this investigation fit into your class time? _____

_____

What changes can you make to improve your planning next time? _____

_____

Guiding and Facilitating Learning

How well did you focus and support inquiry while interacting with students?

_____

What changes can you make to improve classroom management for the next investigation or the next time you teach this investigation? _____

_____

Teacher Commentary

How successful were you in encouraging all students to participate fully in science learning? _____

_____

How did you encourage and model the skills values, and attitudes of scientific inquiry? _____

_____

How did you nurture collaboration among students? _____

_____

Materials and Resources

What challenges did you encounter obtaining or using materials and/or resources needed for the activity? _____

_____

What changes can you make to better obtain and better manage materials and resources next time? _____

_____

Student Evaluation

Describe how you evaluated student progress. What worked well? What needs to be improved? _____

_____

How will you adapt your evaluation methods for next time? _____

_____

Describe how you guided students in self-assessment. _____

_____

Self Evaluation

How would you rate your teaching of this investigation? _____

_____

What advice would you give to a colleague who is planning to teach this investigation? _____

_____

NOTES

Teacher Commentary

INVESTIGATION 3: WEATHER MAPS

Background Information

Weather Maps

Until the advent of computers, up until the middle of the 20th century, weather maps were plotted by hand from weather data gathered over large areas. Nowadays, weather observations are fed into a computer, and the maps are plotted and distributed electronically. Weather maps used by meteorologists contain a lot of information, including not only the basic data taken at a large number of stations but also the results of interpretation on the part of meteorologists. These maps include information about air masses, frontal systems, and high-pressure and low-pressure areas. Weather maps are described by meteorologists as synoptic, meaning that they display weather conditions as they exist at exactly the same time over an area much larger than can be seen from one point on the Earth's surface.

Weather maps that display conditions at higher altitudes are also an important tool for meteorologists. It might seem natural to plot a map that shows the conditions in the atmosphere at a given altitude. What's commonly done, however, is to plot what are called constant-pressure charts, which show conditions at a surface of constant atmospheric pressure; 700 mb (millibars) and 500 mb are commonly used. A bar is a unit of pressure. One bar is very nearly equal to the average sea-level atmospheric pressure. One millibar is one-thousandth of a bar, so the average atmospheric pressure at sea level is about 1000 mb. The altitude where the atmospheric pressure is 700 mb is about 3000 m, and the level where the pressure is 500 mb is about 5000 m.

Constant-pressure surfaces even higher in the atmosphere are also used. A constant-altitude chart would be contoured in terms of pressure, just as surface weather charts are, but a constant-pressure chart is contoured in terms of altitude. The isolines on a constant-pressure chart, then, are curves showing equal altitudes.

Weather maps shown by the media for the general public are much simpler than those used by meteorologists. There was a time when such maps showed isobars (lines connecting points of equal atmospheric pressure), but nowadays most of them show only frontal systems, the locations of high-pressure and low-pressure areas, and areas with precipitation. Your students might be interested in seeing "real" weather maps. Go to the *IES* web site for links that contain weather maps and more information about the weather.

Air Masses

An air mass is a very large mass of air with temperature and moisture content that are similar, at any given altitude, relative to adjacent air masses. An air mass can occupy many thousands of square kilometers, over the land or over the ocean. Although boundaries between air masses are not knife-sharp, they are zones of relatively rapid change in atmospheric characteristics—relative, that is, to the broad areas with very similar characteristics within the interiors of the air masses.

Regions of the Earth where air masses are formed are called source regions. These are places where, owing to the nature of the global general circulation of the atmosphere, the air over large areas moves slowly or is almost stagnant for many days or even a few weeks. The longer the air stays over a given area, the more it takes on the temperature and moisture characteristics of the underlying land or ocean surface. As the general circulation changes, eventually the air mass moves away from its source area. As it moves across distant areas, its characteristics are gradually changed as it gives up or takes on moisture and as it is warmed or cooled. Conversely—and this is what is important for your students—this process changes the weather in the areas over which the air mass is moving.

Weather Fronts
Weather fronts—or just "fronts"—are zones of transition between two air masses with different densities. These different densities are usually caused by differences in temperature. Contrasts in temperature and humidity across fronts can be very large. You're used to seeing fronts represented on surface weather maps in the form of curving lines that separate different air masses. Keep in mind, however, that the zones of transition extend upward from the surface to high altitudes: air masses, and the fronts that bound them, exist in all three dimensions, laterally and vertically. Frontal zones usually slope upward rather than being vertical. Slopes on fronts are usually quite gentle. The cross-sectional diagrams and block diagrams of frontal zones that are commonly shown in meteorology textbooks are usually greatly exaggerated in the vertical scale, in order to show the features of the fronts more clearly. A typical frontal zone might cover 50 or 100 km on the ground while spanning an altitude difference of only 1 km.

Fronts are classified as warm fronts, cold fronts, or stationary fronts in accordance with the relative movement of the adjacent air masses. If colder and denser air is moving in such a way as to displace warmer and less dense air, the front is called a cold front. If warmer and less dense air is moving in such a way as to displace colder and denser air, the front is called a warm front. By the effect of buoyancy, the denser air of the cold air mass forms a lower wedge, and the less dense air of the warm air mass forms an overlying sloping sheet. In either case, whether the cold air is underrunning the warm air or the warm air is overriding the cold air, there is a tendency for the warm air to be lifted. That's the basic reason why precipitation is so commonly associated with frontal systems. If the boundary between the cold air mass and the warm air mass is approximately stationary relative to the underlying ground surface, the front is said to be a stationary front.

More Information...on the Web
Go to the *Investigating Earth Systems* web site www.agiweb.org/ies for links to a variety of web sites that will help you deepen your understanding of content and prepare you to teach this investigation.

Teacher Commentary

Investigation Overview
In **Investigation 3**, students become familiar with the information found on weather maps by studying maps from different newspapers. They complete a simple experiment to help them understand how warm and cold air masses interact. Students also investigate how atmospheric pressure changes with altitude. **Digging Deeper** reviews the history of weather map production, and explains terms typically found on a weather map, including atmospheric pressure, high-pressure and low-pressure areas, and air masses and fronts.

Goals and Objectives
As a result of **Investigation 3**, students will have a greater understanding of the information found on weather maps.

Science Content Objectives
Students will collect evidence that:
1. A great deal of information on weather maps is embodied in symbols.
2. Weather patterns typically move across the United States from west to east.
3. Warmer air masses rise over cooler air masses.
4. Air pressure decreases upward in the atmosphere because at higher levels in the atmosphere there is less air above, and therefore less weight of a unit column of air above the given level.
5. An isotherm is a curving line that connects points on the map where temperatures are the same. An isobar is a line that connects points of equal pressure.

Inquiry Process Skills
Students will:
1. Observe and record their impressions of a phenomenon.
2. Generate questions to answer by inquiry.
3. Devise methods of answering questions using models.
4. Collect evidence from the models.
5. Analyze evidence from the models.
6. Arrive at conclusions based on evidence.

Connections to Standards and Benchmarks
In **Investigation 3**, students examine the information found on weather maps and use models to understand some of the factors that influence weather. These observations start them on the road to understanding the National Science Education Standards and AAAS Benchmarks shown below.

NSES Links

- Different kinds of questions suggest different kinds of scientific investigations. Some investigations involve observing and describing objects, organisms, or events; some involve collecting specimens; some involve experiments; some involve seeking more information; some involve discovery of new objects and phenomena; and some involve making models.

- The atmosphere is a mixture of nitrogen, oxygen, and trace gases that include water vapor. The atmosphere has different characteristics at different altitudes.

- Global patterns of atmospheric movement influence local weather.

AAAS Links

- Because the Earth turns daily on an axis that is tilted relative to the plane of the Earth's yearly orbit around the Sun, sunlight falls more intensely on different parts of the Earth during the year. The difference in heating of the Earth's surface produces the planet's seasons and weather patterns.

- Models are often used to think about processes that happen too slowly, too quickly, or on too small a scale to be observed directly, or that are too vast to be changed deliberately.

Teacher Commentary

Preparation and Materials Needed

Preparation
Part A
You will need to collect copies of weather maps from three different newspaper sources. The maps should represent the same day. A national daily newspaper, like *USA Today*, or a regional newspaper will be a good source to start with. You can also visit the *Investigating Earth Systems* web site to obtain weather maps online. You will also need weather maps with the following features:
- symbols for sky conditions (sunny, partly cloudy, etc.)
- symbols for high and low pressure
- temperatures only
- 10° isotherms

If you are having trouble finding any of these maps, visit the *Investigating Earth Systems* web site for suggestions.

Part B
Aside from collecting the necessary materials, this part of **Investigation 3** requires little advance preparation. You may want to set this experiment up as a station. Students can take turns with the experiment while working on the weather-map investigation. Another option is to do this part of the investigation as a demonstration.

Part C
You will need to obtain an aneroid barometer that has a scale marked in inches of mercury. One for the class should suffice, because these instruments can be expensive and students can take turns reading the barometer.

To complete this part of **Investigation 3**, you will need to take your class to a building that has at least four stories and has an elevator. Students will measure how atmospheric pressure changes as they ride up and down the elevator.

If it is not possible for you to take your class to a tall building with an elevator, you could carry out this activity yourself. You could either videotape the barometer as you ride the elevator or provide students with copies of the data you collect. Another option would be to visit the *Investigating Earth Systems* web site and download data that illustrates how barometric pressure changes with altitude.

Materials
Part A:
- 3 weather maps from different newspapers
- information about weather-map symbols

- weather map with symbols for cloudy skies, partly cloudy skies, and rain*
- weather map with symbols for high-pressure and low-pressure areas*
- overhead transparency sheet
- 2 or 3 colors of overhead transparency markers
- weather map with temperatures only*
- weather map showing 10° isotherms*
- colored pencils

Part B:
- two identical clear plastic 500-mL bottles with medium-sized necks
- caps for the bottles
- one piece of poster board, 10 cm x 10 cm
- supply of hot and cold water
- food coloring: red and blue
- sink or a large pan

Part C:
- aneroid barometer with scale marked in inches of mercury

* The *Investigation Earth Systems* web site provides suggestions for obtaining these resources.

Teacher Commentary

NOTES

Investigating Climate and Weather

INVESTIGATING CLIMATE AND WEATHER

Investigation 3:

Weather Maps

Materials Needed

For this part of the investigation your group will need:

- three weather maps from different newspapers
- information on weather-map symbols
- weather map with symbols for clear skies, cloudy skies, partly cloudy skies, and rain
- weather map with symbols for high pressure and low pressure
- overhead transparency sheet
- two or three colors of overhead transparency markers
- weather map with temperatures only
- weather map showing 10° isotherms
- colored pencils

Key Question

Before you begin, first think about this key question.

What can weather maps tell you about weather?

In a previous investigation you learned about some of the techniques used to measure elements of weather. You have already learned some of the ways this information is displayed on maps. What are some of the other ways this is done?

Share your thinking with others in your group and with your class. Keep a record of the discussion in your journal.

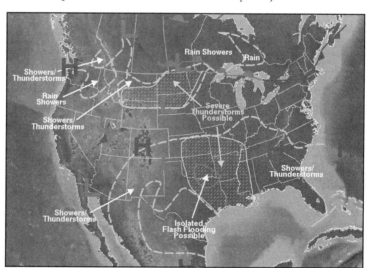

Investigate

Part A: Working with Weather Maps

1. Obtain copies of three newspaper weather maps from different sources.

Investigating Earth Systems

Teacher Commentary

Key Question

Use the **Key Question** as a brief warm-up activity to elicit students' ideas about the information that weather maps can communicate. Write the question on the blackboard or on an overhead transparency. Have students record the question and their answers in their journals. Discuss students' ideas. Ask for a volunteer to record responses on the blackboard or on an overhead transparency. Circulate among the students, encouraging them to copy the notes in an organized way.

Student Conceptions about Weather Maps

Students are likely to have seen weather maps on television and in the newspaper. After completing **Investigation 2**, they should be familiar with some ways in which information is displayed on maps. Ask students to think about this information and recall what they learned in the previous investigation. Ask them if they have ever seen weather maps that contained information not seen in **Investigation 2**; if so, what did they look like and what information did they convey?

Answer for the Teacher Only

Today's weather maps provide a wide variety of information on the weather. These maps can tell you things as simple as "is today a good day to go to the beach or have a picnic, or would tomorrow be better." Beyond that, basic weather maps that show high-pressure and low-pressure systems, temperature, cloud cover, and forecast precipitation can be used to understand the relationship between atmospheric pressure and temperature, wind, rain, clouds, and other more severe weather phenomena. Additionally, weather maps today often can address a wide range of information that includes (among other things) pollen forecasts for people with allergies, travel forecasts for those traveling, etc.

Assessment Tool
Key–Question Evaluation Sheet
Use this evaluation sheet to help students understand and internalize basic expectations for the warm-up activity.

About the Photo
This weather map contains information about precipitation and shows the location of high- and low-pressure systems.

Investigation 3: Weather Maps

In your group, look for at least five things that the maps have in common.

a) Write these down. Share your list with another group.

b) What kinds of information are included in all three weather maps?

c) Make a list of what you already know about the information on weather maps.

d) Write down what you would like to know.

2. Read any information you have available on weather-map symbols.

Share your information about weather-map symbols with the rest of the class.

a) Make up a class chart of weather symbols to use for reference. Include a copy in your journal.

b) How many of your weather-map questions above can you now answer with this information? Record the answers in your journal.

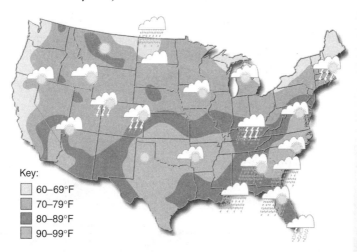

Key:
- 60–69°F
- 70–79°F
- 80–89°F
- 90–99°F

3. Obtain a weather map that has symbols for sky conditions on it.

Place an overhead transparency sheet on top of the map and copy the outline of the map.

Investigating Earth Systems

C 23

Teacher Commentary

Investigate
Teaching Suggestions and Sample Answers

Teaching Tip
Students are again encouraged to generate questions about their observations and consider ways in which they could investigate to find answers. Much of **Investigation 3** involves interpreting maps. Students need to recognize that maps are scientific tools, that they contain scientifically recorded information, and that this mapped information can be interpreted and used by scientists and others.

Part A: Working with Weather Maps
1. Circulate around the room to make sure that your students are not having difficulty. They may need your help in interpreting some of the information contained on weather maps. Most maps include a key. Encourage students to try to figure out the symbols first. Have them consult with you about symbols they do not understand.

   a) Answers will vary. Students may note the information found on the maps, the scale of the map, the presence of a key, etc.

   b) Answers will vary. A sample response might state that all three maps show temperature, warm and cold fronts, areas of high and low pressure, areas of precipitation, and sky conditions.

   c) Answers will vary.

   d) Answers will vary.

2. Although weather reports often share a fairly common set of symbols, there is also variation. Some of the commonly used symbols are shown on page C7 of the student text.

   a) Help your students develop a common set of weather map symbols they can use for their own investigations throughout this module.

   b) Answers will vary, but students should address any questions they posed in **Step 1(d)**.

3. Help students to design a sensible key for their colored map. Ask them to consider carefully which colors they will use for this purpose and why. You may wish to refer students to the map on C23 as a guide. A blackline master of this map is available at the back of this Teacher's Edition.

Investigating Climate and Weather

INVESTIGATING CLIMATE AND WEATHER

On the transparency, circle all the areas that have a sunny symbol. Shade these areas with one color of overhead transparency marker.

Make a key showing which color you are using for sunny areas on the map.

4. Repeat this process using another color for precipitation areas. By convention, green is usually used to indicate precipitation on weather maps. Remember that precipitation includes rain, drizzle, snow, sleet, and hail.

Add this color to your key.

All the uncolored areas will be cloudy or partly cloudy. You can leave these uncolored or use a new color to show them.

5. Put your transparency sheet on a map that shows high-pressure areas, low-pressure areas, and fronts.

a) What relationships can you see between the first map, showing the sky conditions and the second map, showing the pressure systems and fronts? Be sure that you are comfortable with the relationships between sky conditions and pressure systems before you move on. Ask for help as you need it.

Inquiry

Consider Evidence

Scientists look for likely explanations by studying patterns and relationships within evidence. By circling similar types of weather, you developed groupings that reveal patterns. Meteorologists also group different types of weather to reveal patterns that help them correlate factors like high-pressure systems and fair weather.

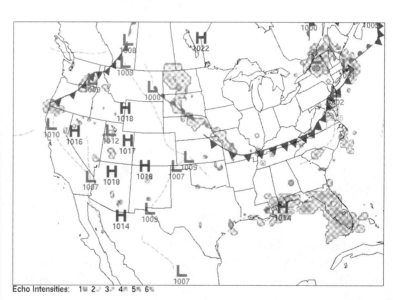

Investigating Earth Systems

Teacher Commentary

4. Help students design a sensible key for their colored map. Ask them to consider carefully which colors they will use for this purpose and why.

Teaching Tip
By circling similar types of weather, students are grouping similar entities. This grouping, in turn, reveals a weather pattern. Meteorologists group different types of weather on maps to reveal these patterns. This helps meteorologists to make correlations between such factors as high-pressure systems and sunny weather.

5. With careful plotting, students should be able to see some patterns and relationships between sky conditions, pressure, and fronts. Help them to clarify their ideas here. You may wish to refer students to the map on C24 as a guide. A blackline master of this map is available at the back of this Teacher's Edition.

    a) This entire exercise should help your students recognize the relationship between fronts, pressure systems, and then the manifestations of weather (clear, sunny, cloudy, foggy, etc.).

Assessment Tool
Journal–Entry Evaluation Sheet
Use this sheet as a general guideline for assessing student journals, adapting it to your classroom if desired.

Journal–Entry Checklist
Use this checklist as a guide for quickly checking the quality and completeness of journal entries.

Investigation 3: Weather Maps

Conduct Investigations

Consider Evidence

6. Obtain a weather map showing temperatures only.

   With different colored pencils, draw smoothly curving lines to connect points on the map where temperatures are the same. These curves are called isotherms (*iso* means same, and *therm* stands for temperature). Use a contour interval of 10°F.

   For example, your map might show an isotherm for 50°F. The curve would pass through points where the temperature was 50°F. Drawing contours is not easy, because most of the temperatures shown on the map are different from the ones you selected for the isotherms. You have to interpolate. To interpolate means to find a value that falls in between two other values. For example, suppose that one point on your map has a temperature of 47°F and a nearby point has a temperature of 55°F. You know that the 50°F isotherm has to run somewhere between those two points. Also, it has to be closer to the 47°F point than the 55°F point. That's because 50° is closer to 47° than to 55°. An example has been provided.

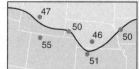

Teacher Commentary

6. Constructing the isotherms may be confusing or frustrating for your students. It's fairly easy when the data points are abundant and when the values vary smoothly over the area. However, when data points are few and there is "scatter" or inconsistencies in the data, drawing isotherms can be very subjective. Obviously, isotherms go right through the "boundary" values. When an isotherm has to fall between two values that are larger and smaller than a "boundary" value, place the isotherm proportionally. For example, in the map on page C26 of the student text, note that, in Wyoming and Colorado, the 50° isotherm is one-third of the way from the 49° value and the 52° value.

Teaching Tip
The map on page C26 of the Student Edition has been reproduced without any isotherms and is included as a **Blackline Master** at the back of this Teacher's Edition. Use this **Blackline Master** to make an overhead that can be used as a tool to demonstrate how to contour isotherms.

Assessment Tool
Journal–Entry Evaluation Sheet
Use this sheet to help students learn the basic expectations for journal entries that feature the write-up of investigations. It provides a variety of criteria that both you and your students can use to ensure that their work meets the highest possible standards and expectations. Adapt this sheet so that it is appropriate for your classroom, or modify the sheet to suit a particular investigation.

INVESTIGATING CLIMATE AND WEATHER

Draw your isotherms at first with light pencil lines, and use an eraser to adjust them if necessary. Once you feel comfortable with your isotherms, draw them permanently with the colored pencils.

7. Now look at a new weather map that just shows bands of temperature readings. The boundaries between these bands are isotherms.

Study the temperature bands on your map carefully. Discuss and answer the following questions:

a) How do the temperature bands help you to understand what weather is like in different parts of the country?

b) How might this be useful in forecasting the weather conditions for a particular area?

c) What patterns can you see in temperatures? (Where is it typically warmer, for example? Where are the temperatures typically cooler?)

d) What reasons can you think of to explain the temperature patterns you notice?

Inquiry

Hypotheses

When you make a prediction and give your reasons for that prediction, you are forming a hypothesis. A hypothesis is a statement of the expected outcome of an experiment or observation, along with an explanation of why this will happen.

A hypothesis is never a guess. It is based on what the scientist already knows about something. A hypothesis is used to design an experiment or observation to find out more about a scientific idea or question.

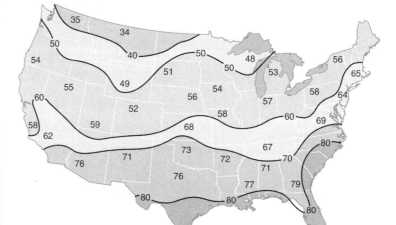

8. Test your knowledge of what a weather map tells you by using a weather map (past or current) to write a weather report for your part of the country.

a) In your journal write the report so that it is easily understandable by others.

Teacher Commentary

7. Students should recognize that temperature bands help to generalize weather across the country and make weather conditions easier to forecast. Students should be able to discover, through the use of the maps, that lower latitudes (those closer to the Equator) tend to have higher temperatures overall, and that higher latitudes tend to have lower temperatures. You could discuss this with your students as an example of an inverse relationship. Another temperature-related inverse relationship is the drop in temperature with altitude (up to a certain point), which students will investigate in upcoming investigations.

Teaching Tip
As students investigate isotherms, ask them if they can think of reasons why the temperature bands do not go straight across the country. If possible, show them a map of the jet stream to help them understand the shape of the isotherms. Visit the *Investigating Earth Systems* to find illustrations of the jet stream.

8. a) Student weather reports will vary. Draw their attention to the increasing complexities of their local self-generated weather report. Ask them to think about what they know now about weather that they did not know when they began the module.

Investigation 3: Weather Maps

Part B: The Movement of Air Masses

1. Read the steps for Part B of the investigation. Although you will use water instead of air, the investigation is a model of what happens when two air masses of different temperature meet.

   a) Before you conduct the investigation, make and record your prediction about what will happen when the card is removed between the two bottles. Also record your reason for your prediction. This is your hypothesis.

2. Fill a 500 mL bottle with cold water and add a few drops of blue food coloring to it.

3. Fill another 500 mL bottle with hot water and add a few drops of red food coloring to it.

   Cap this bottle and shake gently to mix.

   Uncap both bottles.

4. Place a piece of poster board card on top of the cold water bottle.

   Hold the bottle over the sink or the large pan.

   Hold the card tightly to the neck of the bottle, and quickly invert it over the hot water bottle.

   Put the bottles together at their necks. Make sure they match exactly.

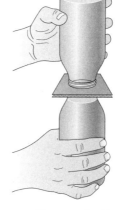

 The water should not be so hot that it can cause burns. Clean up spills.

5. While holding both bottle necks together (cold on top), have someone in your group quickly pull the card out from between the bottles.

   Keep the two necks together and observe what happens to the red and blue water.

   a) Record what you observe.

6. When all water motions seem to be completed in the bottles, empty out the colored water and rinse the bottles for the next group.

Materials Needed

For this part of the investigation your group will need:

- two identical, clear, heavy plastic, 500 mL bottles with medium-size necks
- caps for the two bottles
- one piece of poster board, 10 cm × 10 cm
- supply of hot water and cold water
- blue and red food coloring
- sink or large pan over which the bottles can be inverted

Teacher Commentary

Part B: The Movement of Air Masses

1. Students should read through all of the steps of **Part B** before they begin. You may want to do **Part B** as a demonstration, because students are liable to have difficulty pulling the piece of poster board out from between the bottles, and it could get messy.

   a) Student predictions will vary. It is likely that students will say that the red (warm) and blue (cold) waters will mix to form purple water. (See the sample answer for **Step 6**.)

2 – 5. Students can use the instructions and illustrations in the text to help them set up the experiment. Make sure that the steps are carried out over a sink or large pan.

Investigating Climate and Weather

INVESTIGATING CLIMATE AND WEATHER

Answer the following questions in your journal:

a) What happened when the two bottles were put together? How could you explain this?

b) Air is a fluid, just like water. How does what you observed with the two bottles explain what happens when cold air masses and warm air masses meet?

c) When cold and warm air masses meet, what do you expect to happen? Why is that?

d) How did your results compare with your hypothesis?

Part C: Atmospheric Pressure

Materials Needed

For this part of the investigation your group will need:

- aneroid barometer, with scale marked in inches of mercury

1. Your teacher will arrange a field trip to a nearby building that is at least four stories tall and has an elevator.

   At the lowest floor, before you get in the elevator, read the pressure with the barometer. To do that, tap the glass front very gently with your fingernail several times and watch the position of the needle on the dial.

   a) Record the average position of the needle as you tap the glass.

Stay with an adult supervisor at all times.

Teacher Commentary

6. a) The cold blue water flowed downward into the lower bottle, and at the same time the warm red water flowed upward into the upper bottle. Technically, this is called exchange flow. It happens because the warm water is less dense, and it rises buoyantly upward to displace the denser cold water. The effect is the same as if the lower bottle had been filled only with air, which is far less dense than water.

   b, c) When a cold air mass and a warm air mass converge (i.e., move toward one another), the cold air mass wedges beneath the warm air mass because the cold air is denser. The air, therefore, acts in the same way as the water in the bottles.

   d) Answers will vary.

Part C: Atmospheric Pressure

Teaching Tip
This part of **Investigation 3** is designed to provide students with a hands-on opportunity to see how atmospheric pressure changes with altitude. However, it is unlikely that you will have easy access to a tall building with an elevator. See the **Preparation** section above for some suggested alternatives to completing the experiment. The questions may need to be modified (or omitted) depending on how you choose to deal with the exercise.

1. Gently tapping the glass front of the barometer will help to adjust the position of the needle.

   a) Students should record the position of the needle on the ground floor.

3. a) Students should observe that the barometric pressure decreases as they ride up the elevator.

   b) The reading on the top floor of the building should be less than the reading on the ground floor.

4. a) Again, students should observe that the reading on the top floor of the building should be less than the reading on the ground floor.

   b) Answers will vary. Air pressure decreases upward in the atmosphere because, at higher levels in the atmosphere, there is less air above to cause pressure.

5. Answers to these questions will vary according to the measurements that you took.

Investigating Climate and Weather

Investigation 3: Weather Maps

2. Take the elevator to the top floor, carrying the barometer with you.

   While the elevator is going up, watch the needle. To help the needle adjust its position, you can tap the glass front gently with your fingernail now and then.

3. When the elevator has reached the top floor, step out of the elevator and read the barometer again.

   a) Record how barometric pressure changed as you were riding up the elevator.

   b) Record the barometer reading on the highest floor.

4. Ride the elevator back down to the lowest floor.

   Read the barometer again.

   a) Record this reading, and any relationships between the top and bottom floor readings.

   b) Explain what you think caused the change in the reading of the barometer as you rode up in the elevator.

5. Most barometers record the atmospheric pressure in "inches of mercury." The average atmospheric pressure at sea level (zero elevation of the land surface) is about 30 in.

   Calculate the percentage change in the atmospheric pressure as you went up in your elevator ride.

   a) Find the difference in pressure between the lowest floor and the highest floor.

   b) Divide the difference in pressure by the value for the reading at the bottom floor.

   c) Then multiply the result by 100 to convert your answer to a percentage. This is approximately the percentage of the Earth's atmosphere you went up through on your elevator ride!

6. Look at a weather map that shows barometric pressure across the United States.

   a) Where is the air pressure the lowest? Where is it highest?

   b) What do the H (or High) and L (or Low) symbols indicate on the map?

   c) How does the pressure change moving away from the H? How does the pressure change moving toward the L?

Inquiry
Using Mathematics

In this investigation you made measurements using inches of mercury as a unit to collect data. (In the International System of Units, pressure is measured in pascals, symbol Pa.) Calculations were then used to interpret the data you collected. Scientists also use mathematics in their investigations.

Teacher Commentary

NOTES

INVESTIGATING CLIMATE AND WEATHER

d) What is the general relationship between air pressure (high or low) and the type of weather (storms, sunny, etc.) in a region?

e) In a rapidly ascending elevator or an airplane taking off, you may feel a popping sensation in your ears. What do you suppose causes this, and how might it be related to the drop of air pressure with increasing altitude?

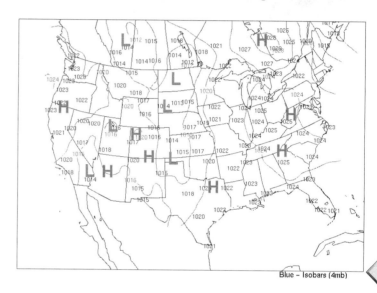

Blue – Isobars (4mb)

As You Read...
Think about:
1. Who drew the first weather map, and how was the information obtained?
2. What is an isotherm? What is an isobar?
3. Why does air pressure always decrease upward in the atmosphere?
4. Why is the weather in high-pressure areas usually fair? Why is the weather in low-pressure areas usually cloudy and stormy?
5. What is the difference between a cold front and a warm front?

Digging Deeper

WEATHER MAPS

A weather map is a graphical model of the state of the atmosphere over a broad region at a specific time. Meteorologists call these synoptic maps (*syn-* means at the same time, and *-optic* stands for seeing). In the late 1700s, Benjamin Franklin drew the first synoptic weather map. He asked a number of friends living in eastern North America to record the weather each day for several days and then mail their journals to him. He then drew weather maps for each day. The mails were very

Teacher Commentary

Teaching Tip
Students can use the map on page C30 of the student text to answer these questions, or you can visit the *Investigating Earth Systems* web site to download a current map that shows barometric pressure. The answers to the questions below reflect the map in the text.

6. a) The pressure is generally lowest in the central plains region (extending northward from northern Texas up through the Dakotas and Montana into Canada) with the lowest pressure occurring in Canada on the Alberta–Saskatchewan border (<1012 mb).

   b) The "H" symbols show the centers of areas of high atmospheric pressure, and the "L" symbols show the centers of areas of low atmospheric pressure. Although it might not occur to your students, it's worth keeping in mind that these pressures are always adjusted to sea level. Atmospheric pressure varies with elevation as well as with the state of the weather. Therefore, in order to compare pressures from place to place, the pressures are cited as if they had been taken at sea level. You could imagine digging or drilling a deep well, open upward to the atmosphere and with its base at sea level, and then taking the pressure reading at the base of the well.

   c) The atmospheric pressure goes down as one moves away from the high pressure region (the "H") and towards the low pressure region (the "L").

   d) In general, high pressure systems tend to be clear and sunny, while low pressure systems tend to be more cloudy and rainy. Student can compare the maps on pages C23 and C24 to determine these relationships if local maps aren't available.

   e) Student responses to what causes the "popping" of their ears may vary, but they should make the correlation that it's the ear's way of responding to changes in pressure. The popping results from equilibration of pressure in the inner part of the ear inside the ear drum (also known as the tympanic membrane) with the changing external pressure.

Digging Deeper
This section provides text, illustrations, and maps that give students greater insight into the components of weather maps. You may wish to assign the **As You Read** questions as homework to help students focus on the major ideas in the text.

As You Read...
1. Benjamin Franklin drew the first weather map by collecting weather journals from friends living in eastern North America.

2. An isotherm is a curving line that connects points on the map where temperatures are the same. An isobar is a line that connects points of equal pressure.

3. Air pressure decreases upward in the atmosphere because at higher levels in the atmosphere there is less air above, and therefore less weight of a unit column of air above the given level.

4. High-pressure areas are generally fair because air in the center of a high tends to be sinking. That happens because air flows outward from the center of high pressure toward surrounding areas of lower pressure. Because of the Coriolis effect, this outward motion takes the form of a spiral, but that need not concern you or your students at this point. Clouds require rising air to form, so high-pressure areas tend to have clear weather. Conversely, low-pressure systems are usually associated with stormy weather because surface winds spiral inward toward the center of low pressure, and as they do, they cause air to rise upward at the center of low pressure, thus producing clouds.

5. A cold front is a band or zone where a cold air mass is wedging under a warm air mass. A warm front is a band or zone where a warm air mass is moving up over a cold air mass.

Assessment Opportunity

You may wish to rephrase selected questions from the **As You Read** section into multiple choice or "true/false" format to use as a quiz. Use this quiz to assess student understanding and as a "motivational tool" to ensure that students complete the reading assignment and comprehend the main ideas.

Teacher Commentary

NOTES

Investigation 3: Weather Maps

slow then, so that Franklin was not able to draw the maps until long after the observations were made. Nonetheless, the maps were useful because they revealed, for the first time, that winds blow around the centers of storm systems that move day by day.

Weather maps that are available to the public vary a lot in how much information they show. Most weather maps in newspapers and on television show only temperature bands that are defined by isotherms, areas of precipitation, the location of high-pressure systems and low-pressure systems, and weather fronts. More specialized weather maps show air pressure, wind speed and direction, cloud cover, and precipitation. The maps you created in Investigation 1 were a simplified version of this kind of weather map.

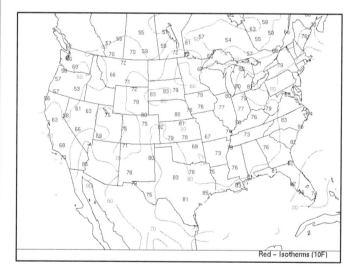

Red – Isotherms (10F)

Atmospheric Pressure

Air has weight. That idea might seem strange to you, because air seems very thin, even at sea level. Remember, however, that the atmosphere extends to great

Teacher Commentary

About the Photo

The map on page C31 of the student text shows temperature and isotherms across the United States, much like the map that students created in **Part A** of **Investigation 3**. You can visit the *Investigating Earth Systems* web site to download other "sophisticated" weather maps of the type used by scientists.

INVESTIGATING CLIMATE AND WEATHER

altitudes. You can think of air pressure as the weight of a column of air above a unit area on the Earth's surface. The column of air above a square area that is one inch on a side averages about 14.7 lb. at sea level. In the metric system, that's about 1.0 kg/cm^2 (square centimeter). If you try to pump the air out of a closed container, the outside air pressure will cause the container to collapse unless the container is very strong. The reason you don't feel air pressure is that the pressure inside your body is equal to the air pressure acting on the outside of your body!

You saw from your elevator ride with the barometer that the air pressure decreases upward in the atmosphere. That's because at higher levels in the atmosphere there is less air above to cause the pressure.

Detailed weather maps show the atmospheric pressure by means of curved lines called isobars. As with an isotherm for temperature, an isobar connects all points with the same atmospheric pressure. Air pressure always decreases with altitude so that the pressure at the land surface is less where the elevation of the surface is high. To remove the effect of elevation on air pressure readings, meteorologists adjust the readings to what they would be if the weather station were at sea level. The adjusted pressure is what you would measure if you could dig a very deep shaft all the way down to sea level and put your barometer at the bottom of the shaft. The adjusted pressure is plotted on weather maps.

High-Pressure Areas and Low-Pressure Areas

Most weather maps show areas, labeled with an **H** (or **High**), where the atmospheric pressure is relatively high, and areas labeled with an **L** (or **Low**) where the atmospheric pressure is relatively low. The isobars around such areas are usually closed curves with the approximate shape of circles. Viewed from above, surface winds blow clockwise (in the Northern Hemisphere) and outward

Teacher Commentary

NOTES

Investigation 3: Weather Maps

about the center of a high as shown in the diagram. As air leaves the high-pressure area, the remaining air sinks slowly downward to take its place. That makes clouds and precipitation scarce, because clouds depend on rising air for condensation of water vapor. High-pressure areas usually are areas of fair, settled weather. Viewed from above, surface winds blow counterclockwise (in the Northern Hemisphere) and inward about the center of a low as shown in the diagram. Converging surface winds cause air to rise, producing clouds. Low-pressure areas tend to be stormy weather systems.

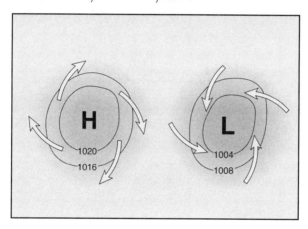

Air Masses and Fronts

Large masses of air, as much as 1000 km across, take on certain weather characteristics when they stay at high latitudes (near the poles) or at low latitudes (near the Equator) for weeks at a time. They may be very cold or very warm, or they may be very humid or very dry. Then, as they move into other areas, they can cause changes in the weather. The coldest winter weather in much of the United States occurs when a bitter cold air mass from the high arctic regions of northeastern Asia, Alaska, or northern Canada sweeps down into the southern parts of North America. At other times, a flow of warm

Teacher Commentary

About the Photo

The diagram on page C33 of the student text illustrates how air circulates around high-pressure and low-pressure areas. In a high-pressure area, surface winds blow clockwise (when viewed from above in the Northern Hemisphere) and away from the center of the high-pressure region (they blow counterclockwise in the Southern Hemisphere). Conversely, in low-pressure systems winds blow counterclockwise (when viewed from above in the Northern Hemisphere) and towards the center of the low-pressure system.

INVESTIGATING CLIMATE AND WEATHER

and humid air from the tropics causes uncomfortably muggy weather over the eastern United States.

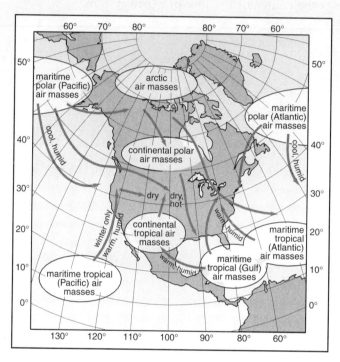

The boundaries between air masses are often zones of very rapid changes in temperature and humidity. Storms (low-pressure systems) tend to develop along these zones of rapid change. The line with the triangular teeth, called a cold front, shows where the cold air mass is wedging under the warm air mass. (See the cross-sectional diagram on the following page.) As the warm air is lifted along the front, thunderstorms may develop. The line in the diagram with the circular teeth, called a warm front, shows where the warm air mass is moving up over the cold air mass. Broad areas of light to moderate rain or snow are often associated with the warm front.

Teacher Commentary

About the Photo

Use **Blackline Master** *Climate and Weather* 3.4, which is available at the back of this Teacher's Edition, to make an overhead of the map on page C34 of the student text. Incorporate this overhead into a lecture or discussion on air masses in North America. An air mass is a large body of air defined by its temperature and moisture content. The type of air mass is determined ultimately by its source region, because that is where the air acquires its basic temperature and humidity properties. In general, air masses that originate over continents are drier than those that originate over the ocean, and air masses that originate at high latitudes are colder than those that originate at low latitudes. Air masses transport heat and moisture in the atmosphere around the globe.

Investigation 3: Weather Maps

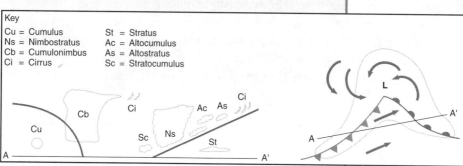

A typical large rainstorm along a zone of rapid change.

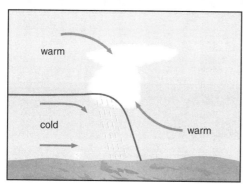
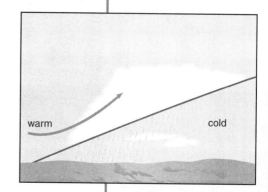

Cross-sections through warm fronts and cold fronts.

If the pattern of the fronts in the diagram looks to you a bit like a wave, you are right. A low-pressure system is developing like a wave along the boundary between the two air masses! You can see that the pattern of surface winds, shown by the arrows, is developing a counterclockwise and inward pattern. After a storm like this develops, the swirling pattern of winds continues for many days, until it finally dies out. A storm of this kind can move hundreds of kilometers in a day, or it can remain in almost the same position for a day or even longer. Much like a Frisbee, the storm combines rotational and translational motion as it travels across the nation.

Teacher Commentary

About the Photo

The top diagram illustrates both a map view (right) and associated cross-sectional view (left) of a large rainstorm along a zone of rapid change. The cross-sectional view depicts the cross section of the line A–A' which is shown in the map view diagram (right). The bottom two panels illustrate cross-sectional views through warm and cold fronts.

When two air masses meet, a boundary (called a front) forms between them. There is very little mixing across the boundary; therefore, when one air mass advances on another, one of the air masses must be displaced. If the advancing air mass is colder, it pushes underneath the warmer mass, and a cold front develops, as shown. This forces the warmer air upward, which often causes clouds and precipitation to form. If a warm air mass advances on a cold air mass, a warm front develops, and the warmer air mass rides up over the colder mass. As the warm air rises, it cools and tends to lose some of its moisture as precipitation.

INVESTIGATING CLIMATE AND WEATHER

Review and Reflect

Review

1. What kinds of information were plotted on the weather maps that you investigated?

2. Use your observations of the two bottles of water in Part B of this investigation to explain what happens when a cold air mass meets a warm air mass. Which air mass is likely to rise? Which air mass is likely to stay near Earth's surface?

3. What evidence did you obtain from your investigation that shows that air pressure decreases with altitude?

Reflect

4. What do weather maps tell you about the weather now and in the future?

5. What evidence of relationships were you able to find between cloud patterns and precipitation?

6. What evidence of relationships were you able to find between cloud patterns and high- and low-pressure systems?

Thinking about the Earth System

7. On your *Earth System Connection* sheet, note how the things you learned in this investigation connect to the different Earth systems.

   a) Describe how air pressure (atmosphere) is related to elevation (geosphere).

   b) Describe how air masses (atmosphere) are related to the regions over which they form (geosphere or hydrosphere).

Thinking about Scientific Inquiry

8. What weather data did you use to look for relationships in this investigation?

9. How did you use mathematics in this investigation?

10. Give an example of how you used evidence to develop ideas in your investigations into weather.

Teacher Commentary

Review and Reflect

Review

Give your students ample time to review what they have learned in **Investigation 3**. In student answers, look for evidence of an understanding of the processes involved, as well as for any misconceptions that still remain. Encourage students to express their ideas clearly and to use correct science terminology where appropriate.

1. Answers will vary depending upon the maps that were viewed, but they are likely to include temperature, air pressure, location of fronts, precipitation, etc.

2. When a warm air mass and a cold air mass meet, the warm air mass rises while the cold air mass stays near the Earth's surface. Encourage students to indicate how the experiment they completed in **Part B** of **Investigation 3** supports this idea.

3. In **Part C** of **Investigation 3**, students should have observed that air pressure decreased as they rode up the elevator.

Reflect

Give students time to reflect on the nature of the evidence they have generated from their investigations. Again, help them to see that evidence is crucial in scientific inquiry.

4. Weather maps can tell current weather conditions, like temperature or where it is raining. When the movements of fronts, air masses, and areas of precipitation are projected or extrapolated into the future, weather maps can also be used to show what weather conditions will be like in the future. Keep in mind, however, that weather prediction involves much more than just simple extrapolation of that kind, because the pattern and nature of weather systems changes with time in complex ways.

5. Obviously, when the skies are clear, there is no precipitation. Precipitation is most often associated with cumulonimbus, nimbostratus, and stratus clouds. Most of the other kinds of clouds usually do not produce precipitation.

6. Clouds are typically associated with low-pressure areas, because converging surface winds around low-pressure areas cause air to rise. Water vapor in the rising air cools and condenses, producing clouds.

Thinking About the Earth System

Give students time to make any connections they can between the Earth System and their exploration of weather mapping, atmospheric pressure, high-pressure and low-pressure areas, and air masses and fronts in **Investigation 3**. Refer students to the diagram on page Cviii of the student text as a means of exploring the Earth System further. Ensure that students add their ideas to their *Earth System Connection* sheets.

7. a) Atmospheric pressure decreases with increasing elevation.

   b) The initial temperature and amount of moisture associated with an air mass is determined by the conditions associated with the land or water mass over which the air mass forms.

Thinking about Scientific Inquiry

In **Investigation 3**, students have been exposed to the idea of forming and testing a hypothesis. Refer students to the first **Inquiry Process** listed on page Cx of the student text: "Explore questions to answer by inquiry." It is important that you allow them time to see how this approach to experimentation fits with scientific inquiry, and the inquiry processes they have been using.

8. Students looked at weather maps that contained information on temperature, precipitation, sky conditions, etc. Students also looked at data on barometric pressure.

9. Students used math to calculate the percentage change in the atmospheric pressure as they rode up the elevator.

10. Answers will vary but can include the weather maps examined during the investigation, the experiment using the hot and cold water to understand how air masses of different temperatures interact, or the measurement of pressure in an elevator to understand how atmospheric pressure varies with altitude.

> **Assessment Tool**
>
> **Review and Reflect Journal–Entry Evaluation Sheet**
> Use the general criteria on this evaluation sheet for assessing content and thoroughness of student work. Adapt and modify the sheet to meet your needs. Consider involving students in selecting and modifying the criteria for evaluating their reflections on **Investigation 3**.

Teacher Commentary

NOTES

Teacher Review

Use this section to reflect on and review the investigation. Keep in mind that your notes here are likely to be especially helpful when you teach this investigation again. Questions listed here are examples only.

Student Achievement

What evidence do you have that all students have met the science content objectives?

Are there any students who need more help in reaching these objectives? If so, how can you provide this?

What evidence do you have that all students have demonstrated their understanding of the inquiry processes?

Which of these inquiry objectives do your students need to improve upon in future investigations?

What evidence do the journal entries contain about what your students learned from this investigation?

Planning

How well did this investigation fit into your class time?

What changes can you make to improve your planning next time?

Guiding and Facilitating Learning

How well did you focus and support inquiry while interacting with students?

What changes can you make to improve classroom management for the next investigation or the next time you teach this investigation?

Teacher Commentary

How successful were you in encouraging all students to participate fully in science learning? _____

_____

How did you encourage and model the skills values, and attitudes of scientific inquiry? _____

_____

How did you nurture collaboration among students? _____

_____

Materials and Resources
What challenges did you encounter obtaining or using materials and/or resources needed for the activity? _____

_____

What changes can you make to better obtain and better manage materials and resources next time? _____

_____

Student Evaluation
Describe how you evaluated student progress. What worked well? What needs to be improved? _____

_____

How will you adapt your evaluation methods for next time? _____

_____

Describe how you guided students in self-assessment. _____

_____

Self Evaluation
How would you rate your teaching of this investigation? _____

_____

What advice would you give to a colleague who is planning to teach this investigation? _____

_____

NOTES

Teacher Commentary

INVESTIGATION 4: WEATHER RADIOSONDES, SATELLITES, AND RADAR

Background Information

Radiosondes

Radiosondes are small boxes that contain instruments and a radio transmitter. Carried upward by balloons, radiosondes measure the vertical distribution of temperature, pressure, and humidity up to an altitude of about 30 km, far above the top of the troposphere (the lowermost layer of the atmosphere, where almost all of what we think of as weather happens).

Temperature is usually measured with an instrument called a thermistor, located just outside the box. Thermistors work on the principle that the electrical resistance of certain materials varies in a known way with temperature. An electric current is passed through the material; the resistance is measured and then correlated with the temperature.

Humidity is measured in a similar way. A small aneroid barometer inside the box measures air pressure. The measurements are transmitted to the surface, and a computer transforms the data into values of temperature, pressure, and humidity. Sometimes, information on wind speed and direction as a function of altitude is measured by tracking the radiosonde visually with a kind of optical instrument called a theodolite, similar to a surveyor's transit. These measurements of wind can be made using just the pilot balloon, without the radiosonde package.

Eventually the balloon carrying the radiosonde bursts, and the radiosonde falls back to Earth, slowed by a miniature parachute. Some of the radiosondes are retrieved in working condition, but others are lost or damaged. Use of radiosondes, especially a sufficiently dense network of them, is an expensive undertaking. However, it is the best way to make systematic measurements of weather conditions in the upper atmosphere.

Weather Satellites

The development of satellites in recent decades for remote sensing of cloud cover has revolutionized weather observations. Weather satellites provide photographs of clouds over the large areas of the Earth's surface where ground-based stations are scarce, especially over the ocean. Information on both cloud height and cloud thickness, as well as cloud type, can be read from visible-light satellite cloud photographs. Thick clouds reflect more visible light than thin clouds, so they appear brighter. Infrared photographs reveal the temperature of the cloud tops. Because it varies generally with height, cloud temperature gives an approximate indication of cloud height.

Two kinds of satellite orbit are useful in making weather observations. Geostationary satellites are those that orbit the Earth at a height such that their speed is the same as the speed of rotation of the Earth's surface. To an observer at the surface, a geostationary satellite stays in the same position in the sky. Geostationary satellites are used for continuous monitoring of a specific region of the Earth. Because their orbits are about 36,000 km above the surface, they can see almost one-third of the Earth's surface in a single photograph—although at the edge of the field of view, the angle of view becomes very oblique. Geostationary satellites transmit their measurements in real time, rather than with a time delay.

There are also satellites that have orbits adjusted to be parallel to the Earth's north–south meridian lines. As the Earth rotates beneath the satellite, each north–south pass of the satellite monitors a different north–south segment of the surface. Eventually the satellite covers the entire surface of the Earth. The advantages of polar-orbiting satellites are that the orbit is lower—at about 850 km—so the photographs are more detailed, and the view is always nearly straight down. The disadvantage of polar-orbiting satellites, however, is that the measurements must be stored for transmission at a later time.

Weather Radar
Weather radar is a remote-sensing instrument that provides meteorologists with a unique and valuable perspective as they monitor the atmosphere. Traditionally, meteorologists have depended on observational data acquired by weather instruments at discrete weather stations. Unfortunately, these stations typically are hundreds of kilometers apart and may not detect small-scale weather systems like thunderstorms unless such systems pass directly over or within sight of the station. Weather radar largely fills in the gaps between weather stations by continuously scanning a broad volume of the atmosphere. Weather radar can locate small, isolated areas of precipitation, track the movement of thunderstorms, distinguish between rain and hail, and provide early warning of thunderstorms with the potential to produce tornadoes or flash floods. The present generation of weather radar can also estimate cumulative rainfall over a broad area, offering a significant advantage over the network of rain gauges, which generally do poorly in representing the great spatial variability of precipitation typical of thunderstorm weather.

The principle of radar is simple. It uses pulses of long-wavelength electromagnetic radiation, termed microwave radiation, with a typical wavelength of 1 to 20 cm. This is very long compared to the wavelength of visible light, which ranges from about 4×10^{-7} m to about 8×10^{-7} m. When the waves encounter solid objects in the atmosphere, some of the wave energy is scattered. Scattering is an effect whereby the electromagnetic waves interact with the solid objects in such a way that some fraction of the waves are redirected in all directions away from the original straight-line path. A small fraction of the wave energy is scattered back in the direction of the source. The effect is different from actual reflection, but the result is similar. Objects sensed by radar can be raindrops, snow crystals, hailstones, and even swarms of insects. The return signal is amplified and processed electronically to be displayed as an image, called an echo, on a screen. The brightness of the echo depends in part upon the density of the objects that are scattering the waves. The radar screen thus shows not only where the precipitation is occurring but also its intensity. At microwave wavelengths of about 1 cm, even cloud droplets can be detected by radar.

A special form of radar, called Doppler radar, can measure the speed at which precipitation is moving horizontally relative to the radar antenna. Snowflakes, raindrops, and hailstones move with the wind as they fall, because they are embedded in the surrounding air. Because of that, Doppler radar reveals wind patterns in areas where precipitation is falling. Doppler radar is especially useful in studying small-scale

Teacher Commentary

weather systems like thunderstorms. To get an accurate picture of wind directions, however, two Doppler radars directed at right angles to one another and some distance apart are needed.

The principle behind Doppler radar is the same as what in physics is called the red shift: the frequency of sound from a source moving toward you is higher than the frequency of sound made by the same source when it is moving away from you. This is because in the former case, the speed of approach is added to the speed of the waves, making the frequency seem greater, whereas in the latter case, the speed of retreat is subtracted from the speed of the waves, making the frequency seem smaller.

More Information…on the Web

Go to the *Investigating Earth Systems* web site www.agiweb.org/ies for links to a variety of web sites that will help you deepen your understanding of content and prepare you to teach this investigation.

Investigation Overview

Students begin **Investigation 4** by plotting data collected from two radiosondes to understand how air temperature changes with altitude. Then, they study the relationship between satellite images and weather maps. Students compare satellite and radar images to understand what information is found on each. **Digging Deeper** explains how scientists use radiosondes to study the weather in the upper atmosphere, and reviews the development and use of weather radar and weather satellites.

Goals and Objectives

As a result of **Investigation 4,** students will develop a better understanding of the various ways in which weather observations are made in the upper atmosphere.

Science Content Objectives

Students will collect evidence that:
1. Satellite images and radar images, as well as other sources of information, are used to make weather maps.
2. Radar is used to detect the intensity of precipitation, and to tell frozen forms from unfrozen forms of precipitation.
3. Weather satellites are valuable tools for monitoring changes in the Earth system, like the development and scale of weather systems.
4. Air temperature decreases with altitude.

Inquiry Process Skills

Students will:
1. Construct a graph from a set of data.
2. Use the data and graph to investigate relationships.
3. Search for patterns and relationships using different representations of weather (maps and satellite images).
4. Arrive at conclusions based on evidence.

Connections to Standards and Benchmarks

In **Investigation 4**, students study the relationship between weather maps, satellite images, and radar images. These observations will start them on the road to understanding the National Science Education Standards and AAAS Benchmark shown below.

NSES Links

- Different kinds of questions suggest different kinds of scientific investigations. Some investigations involve observing and describing objects, organisms, or events; some involve collecting specimens; some involve experiments; some involve seeking more information; some involve discovery of new objects and phenomena; and some involve making models.

Teacher Commentary

- The atmosphere is a mixture of nitrogen, oxygen, and trace gases that include water vapor. The atmosphere has different characteristics at different altitudes.

AAAS Link

- Because the Earth turns daily on an axis that is tilted relative to the plane of the Earth's yearly orbit around the Sun, sunlight falls more intensely on different parts of the Earth during the year. The difference in heating of the Earth's surface produces the planet's seasons and weather patterns.

Preparation and Materials Needed

Part B of **Investigation 4** requires students to observe satellite images and weather maps over a period of days. These images can be obtained on the **Investigating Earth Systems** web site. If you do not have access to the Internet in your classroom, you can print out the necessary maps and images and distribute them to your students. Days should be consecutive. You could also have your students watch television weather reports for homework and report their findings in class. Another option is for you to record television weather broadcasts that show satellite images and then show the recordings to class.

Materials
Part A:
- two sheets of graph paper

Part B:
- satellite images*
- weather maps*
- recordings of television weather broadcasts (optional)

Part C:
no additional materials needed

* The *Investigation Earth Systems* web site provides suggestions for obtaining these resources.

Teacher Commentary

NOTES

Investigation 4: Weather Radiosondes, Satellites, and Radar

Investigation 4:

Weather Radiosondes, Satellites, and Radar

Key Question
Before you begin, first think about this key question.

In addition to weather observations on the Earth's surface, how else is weather data collected?

In the previous investigation, you studied weather maps. Some of the information used to make the maps is obtained at weather stations at the surface. Much of the information, however, is obtained from other sources. How else can weather observations be made?

Share your thinking with others in your group and with your class. Keep a record of the discussion in your journal.

Materials Needed

For this investigation your group will need:

- two pieces of graph paper

Investigate

Part A: Data from Radiosondes

1. The table on the following page gives data on how temperature changes with altitude.

Teacher Commentary

Key Question

Write the **Key Question** on the blackboard or on an overhead transparency. Tell students to write the question in their journals and to think about and answer the questions individually. Tell them to write as much as they know and to provide as much detail as possible in their responses. Remind them that the date and the prompt (question, heading, etc.) should be included in all of their journal entries.

Assessment Tool
Key–Question Evaluation Sheet
Use this evaluation sheet to help students understand and internalize basic expectations for the warm-up activity.

Student Conceptions about Weather Observation in the Upper Atmosphere

Most likely, students do not know how weather in the upper atmosphere is studied, although some students may have heard of weather balloons. They may have some idea that weather can be studied using weather satellites, and they may have seen some satellite images on television weather reports. They may, however, confuse satellite images with those produced by radar. Students may know less about how satellites are deployed around the Earth for this purpose, and how they communicate their images from space.

Answers for the Teacher Only

Radiosondes are instrument packages attached to balloons. They record temperature and pressure as they rise through the atmosphere. Various techniques of remote sensing are now used to gather weather data. The most important ground-based technique is weather radar. Radar images show areas of precipitation. A special radar technique, called Doppler radar, gives the speeds of motion of areas of precipitation. Satellite images show the pattern of weather over large areas of the Earth. Successive satellite images show graphically how weather systems move.

INVESTIGATING CLIMATE AND WEATHER

Radiosonde Data June 26, 2001

Jacksonville, Florida		Fairbanks, Alaska	
Temperature (°C)	Altitude (m)	Temperature (°C)	Altitude (m)
20.6	9	17	138
24.2	88	16.4	197
24.6	203	14.1	610
24.8	327	12.8	842
23.2	610	9.7	1219
21.6	884	7	1545
18.4	1219	3	2164
13.2	1829	−2.5	2923
11.2	2134	−6.4	3658
5	3224	−11.4	4540
−0.3	4267	−16.3	5180
−4.3	4997	−22.1	5970
−12.7	6096	−27.7	6746
−23.7	7570	−32.7	7620
−28.3	8230	−38.7	8469
−39.1	9610	−44.9	9300
−43.7	10830	−54.7	10668
−56.7	11983	−57.7	11900
−59.7	13766	−51.7	12719
−64.1	15240	−50.3	15240
−64.9	16540	−49.3	16619
−67.7	17692	−50.1	18601
−64.9	18700	−48.1	20950
−59.7	20780	−48.6	22555
−54.2	22555	−46.6	23774
−49.8	26518	−43.6	25603
−50.7	27432	−40	27432
−42.1	30480	−36.3	29403
−39.9	31270	−33.3	31840
−39	33223	−28.1	33528

A radiosonde is an instrument package that is carried upward by a balloon. As it rises to great altitudes it makes weather observations.

Teacher Commentary

NOTES

Investigation 4: Weather Radiosondes, Satellites, and Radar

2. For each data set, use a sheet of graph paper to plot how temperature changes with altitude.

   a) Use the vertical axis for altitude above sea level. Use the horizontal axis for temperature. See the sample shown.

   b) Plot the temperature at each altitude.

   c) Then connect the points (which scientists call "data points") with a continuous line. A line like this is called a sounding.

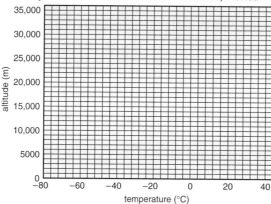

3. Use your graph to answer the following questions:

   a) Does the air temperature generally increase or decrease with altitude?

   b) What do you think is the cause of the increase or decrease?

   c) The "cruising altitude" of commercial jetliners is usually in the range of 10,000 m (about 30,000 ft.) to 13,000 m (about 40,000 ft.). Judging from the two temperature profiles you plotted, what is the typical air temperature in that range of altitudes?

   d) In the Jacksonville sounding, the temperature shows an increase with altitude in the lower part of the atmosphere. What do you think is the cause of that increase? (Hint: the lower atmosphere is heated and cooled from below.)

Inquiry

Mathematical Relationships

In the previous investigation you discovered that the higher the altitude, the lower the air pressure. This is an example of an inverse relationship. In this investigation you are studying another inverse relationship.

Investigating Earth Systems

C 39

Teacher Commentary

Investigate

Teaching Suggestions and Sample Answers
Part A: Data from Radiosondes

1. The radiosonde data were collected at the same time on June 26, 2001, in Jacksonville, Florida, and in Fairbanks, Alaska.

   Teaching Tip
   You may wish to have your students create their graphs using a spreadsheet program like Microsoft Excel. One important science inquiry skill you can remind your students about is the collection and review of data using tools. The computer is a data management tool.

2. Students can use the sample graph on page C39 of the student text as a guide to setting up their own graphs. Remind them to label the axes and include a title for each graph. Final graphs are shown below.

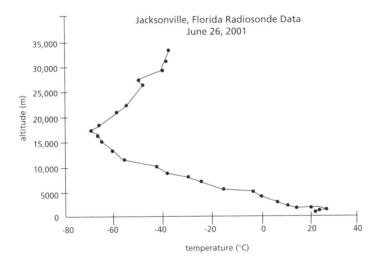

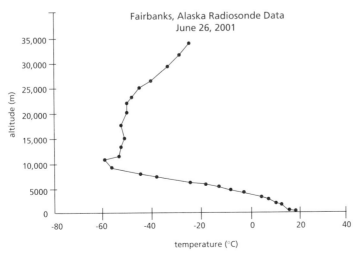

Investigating Climate and Weather – Investigation 4

3. a) Student graphs should show that temperature generally decreases with altitude to a certain point (between about 12,000 and 18,000 m in the data given), and then increases at altitudes above that point.

   b) Student answers to this question will vary, and will likely not include the level of detail presented here. Several factors contribute to the decrease in temperature with altitude. First, air near the ground is warmed by the ground, which absorbs heat from the Sun and radiates it back to the air. This warming decreases with increasing altitude. Additionally, higher up in the atmosphere, air loses more heat to space than it absorbs from sunlight. Another effect that contributes to the general decrease in temperature with altitude is that when a mass of air rises, it expands; the expansion causes the temperature to decrease because the gas molecules have lost some of their heat energy through the work they have done on their surroundings. The high-altitude zone of increasing temperature with increasing altitude, as reflected in the radiosonde data, is related to the absorption of ultraviolet radiation by ozone within the stratosphere.

> **Teaching Tip**
> The atmosphere can be divided into several different layers that have boundaries defined by abrupt changes in temperature. These layers, in order of increasing altitude, are thy troposphere, the stratosphere, the mesosphere and the thermosphere. Temperature generally decreases with increasing altitude in the troposphere and mesosphere, but in the stratosphere, temperature increases with increasing altitude. This temperature increase is related to the absorption of ultraviolet radiation by the ozone layer, which resides in the stratosphere. In the outermost layer of the atmosphere, the thermosphere, solar radiation causes this layer to warm to extremely high temperatures that can exceed 1000°C!

   c) The two temperature profiles indicate an air temperature range of approximately -40°C to -55°C for the cruising altitude of commercial jetliners.

   d) At certain times, especially on clear nights, the ground surface is cooled as it loses heat by radiation to outer space. When that happens, the air temperature near the ground becomes lower than the air temperature higher in the atmosphere. The effect is called an inversion. There are other situations in which inversions develop as well.

Teacher Commentary

Teaching Tip
You can print out data collected using a radiosonde that was launched close to your community by visiting the *Investigating Earth Systems* web site. Students can compare data from their community to data collected from around the world.

Making Connections...*with mathematics*
Graphing and data analysis are both science and mathematics skills. You may want to work in collaboration with the mathematics teacher for your class to maximize these learning skills.

INVESTIGATING CLIMATE AND WEATHER

Part B: Satellite Images

1. Look carefully at the satellite image.

   a) What do you think it shows? How can you check your ideas?

   b) How might satellite images be helpful in forecasting or understanding weather? What do you think?

   c) Compare your ideas with those from another group. How do they compare?

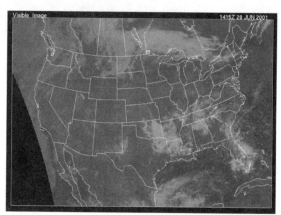

2. Compare the satellite image to the sky conditions shown on a weather map for the same time period.

   a) List and describe as many relationships as you can between the satellite image and the weather elements on your map. Discuss the relationships you discovered with another group.

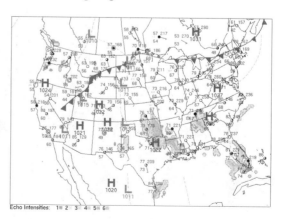

Teacher Commentary

Part B: Satellite Images

1. a) The satellite image on page C40 of the student text shows the location of clouds. Student suggestions on how to check their ideas will vary but could include comparing the satellite image to a weather map or going outside and looking for clouds.

    b) Satellite images taken over time can show how storm systems are moving.

    c) Have students share their ideas about what they think the satellite image shows and how it might be useful in understanding weather.

    ### Teaching Tip
    One way to show students satellite images is to connect a computer that has Internet access to a projector and use it as a visual teaching tool. Visit the *Investigating Earth Systems* web site to find useful links to satellite images.

2. You can have your students use the satellite image and weather map in the student text, or you can provide your own.

    a) Several relationships can be seen when comparing the satellite image to the weather map. Cloud cover is seen associated with the all of the areas of precipitation shown on the weather map (e.g., Florida, Oklahoma, Louisiana, Georgia, Tennessee). Also, there is a lot of cloud cover just north of (and associated with) the long frontal system that extends from New England through Montana. The cloud cover even dips south at the same place that the frontal system shown on the map dips south (North Dakota). Additionally, regions of high pressure (e.g., in the Southwest) have little if any clouds that are visible on the satellite image.

Investigation 4: Weather Radiosondes, Satellites, and Radar

3. Over the next few days, visit a web site that has both satellite images and weather maps. You might also watch television weather reports that show both weather maps and satellite images.

   a) Do you see any pattern in the way weather systems move? Explain.

   b) Compare the movement of weather systems and the map of the movement of air masses (page C34). What relationships can you see?

Inquiry

Using Evidence Collected by Others

In this investigation you used evidence that you were provided to formulate your ideas about weather systems. Meteorologists must also consider the evidence provided by others to develop their ideas about weather patterns.

Part C: Radar Images

1. Compare the radar and satellite images shown on the following page.

   a) What do you notice about the relationship between the clouds and radar echoes?

   b) Are all the clouds producing precipitation? How can you tell?

Teacher Commentary

3. You can visit the *Investigating Earth Systems* web site to find satellite images and maps. If you do not have Internet access for your students, you can have your students watch weather reports as homework or you can record television weather reports and show them to your class.

   a) Answers will vary, but most likely students will find that weather systems typically move across the United States from west to east.

   b) Students should notice that, in general, weather systems and air masses move in the same direction: from west to east.

Part C: Radar Images

Teaching Tip
Make an overhead of each image on pages C42 and C43 of the student text. The images can then be overlaid and projected so students can see how they match up. The images can also be projected side by side for comparison. Copies of each of these images have been provided as **Blackline Masters** and can be found at the back of this Teacher's Edition.

1. a) In general, clouds and radar echoes occur in the same places.

   b) Some (though not all) of the clouds are producing precipitation, as indicated by the corresponding location of radar echoes.

INVESTIGATING CLIMATE AND WEATHER

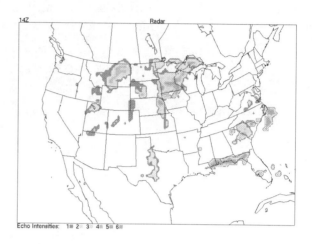

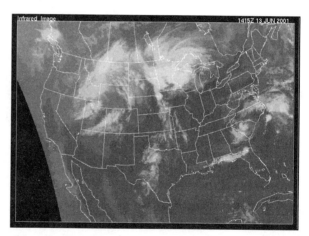

2. Look at the radar image at the top of the next page. It was taken four hours later.

   a) What has happened to the area coverage and intensity of the precipitation over Minnesota and Wisconsin?

   b) What has happened to the area coverage and intensity of the precipitation over North and South Carolina?

3. Look at the infrared satellite image on the next page taken four hours later.

   a) What has happened to the clouds along the border between West Virginia and Virginia?

   b) What do you think is responsible for this change?

Teacher Commentary

2. a) After four hours, the area coverage and intensity of precipitation over Minnesota and Wisconsin have decreased.

   b) After four hours, the area coverage and intensity of precipitation over North Carolina and South Carolina have increased.

3. a) Greater cloud cover and precipitation developed over the West Virginia–Virginia border, and the cloud cover in the region generally moved east during the 4 hours represented by the radar and satellite images.

   b) The movement of a frontal system into the area caused this change.

Assessment Tool
Investigation Journal–Entry Evaluation Sheet
Use this sheet to help students learn the basic expectations for journal entries that feature the write-up of investigations. It provides a variety of criteria that both you and your students can use to ensure that their work meets the highest possible standards and expectations. Adapt this sheet so that it is appropriate for your classroom, or modify the sheet to suit a particular investigation.

Investigating Climate and Weather

Investigation 4: Weather Radiosondes, Satellites, and Radar

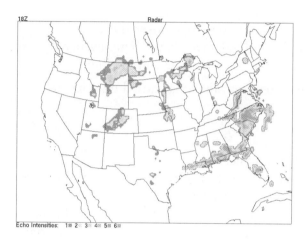

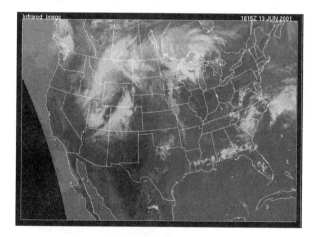

4. Note that on the infrared satellite image, temperature is given on a gray scale—ranging from bright white for lowest temperatures (high clouds) to black for highest temperatures (land). The temperature decreases with altitude, so that high clouds are cold and appear brighter than warmer low clouds.

a) What do the bright white in the clouds over West Virginia and Virginia indicate?

b) Thunderstorm cloud tops rise to great heights. What kind of weather do you think that this area of West Virginia and Virginia was having at the time?

Teacher Commentary

4. a) The bright white clouds over West Virginia and Virginia probably indicate local areas of heavy precipitation.

   b) It is possible that the area of West Virginia and Virginia was having thunderstorms when the satellite image was taken.

Assessment Tools

Journal–Entry Evaluation Sheet
Use this sheet as a general guideline for assessing student journals, adapting it to your classroom if desired.

Journal–Entry Checklist
Use this checklist as a guide for quickly checking the quality and completeness of journal entries.

Group Participation Evaluation Forms I and II
One of the challenges to assessing students who work in collaborative teams is assessing group participation. Students need to know that each group member must pull his or her weight. As a component of a complete assessment system, especially in a collaborative learning environment, it is often helpful to engage students in a self-assessment of their participation in a group. Knowing that their contributions to the group will be evaluated provides an additional motivational tool to keep students constructively engaged.

Group Participation Evaluation Forms I and II provide students with an opportunity to assess group participation. In no case should the results of this evaluation be used as the sole source of assessment data. Rather, it is better to assign a weight to the results of this evaluation and factor it in with other sources of assessment data. If you have not done this before, you may be surprised to find how honestly students will critique their own work, often more intensely than you might do.

INVESTIGATING CLIMATE AND WEATHER

As You Read...
Think about:
1. Why does temperature usually decrease with altitude in the lower portion of the atmosphere?
2. What weather observations are typically made by radiosondes?
3. What piece of weather information are satellites especially good at showing?
4. How is information from radiosondes and satellites transmitted to the Earth?

Digging Deeper

THE WEATHER HIGH IN THE ATMOSPHERE
Radiosondes

You experience weather at the Earth's surface, but there is weather high in the atmosphere also. What is the weather like at high altitudes? Have you ever taken a ride in a hot-air balloon or climbed a mountain? You would know that the air temperature usually decreases with altitude. The basic reason has to do with how the atmosphere receives and loses its heat energy. The Earth's surface is heated by the Sun at some times and places, that is, when and where it is daylight. It loses heat to outer space at all times (day and night) everywhere. On a global and average annual basis, however, the Earth's surface gains more heat than it loses. The atmosphere near the ground is then heated by the ground. High up in the atmosphere, however, the air loses more heat to space than it absorbs from sunlight. Hence, air usually gets colder with increasing altitude—at least up to an altitude of about 10,000 m (33,000 ft.).

Teacher Commentary

Digging Deeper
This section provides text and photographs that give students greater insight into the use of radiosondes, radar, and satellites to measure elements of weather high in the atmosphere. You may wish to assign the **As You Read** questions as homework to help students focus on the major ideas in the text.

As You Read...
1. Temperature usually decreases with altitude in the lower atmosphere, because most heating from the Sun happens at the ground surface rather than in the atmosphere. Higher in the atmosphere, air loses more heat to space than it absorbs.

2. Radiosondes typically make observations of temperature, pressure, and humidity. Tracking of radiosonde movement also yields information on wind speed and direction.

3. Satellites are especially good at showing cloud cover and temperatures of clouds and other surfaces in the sensor's field of view.

4. Radiosondes send weather measurements by radio. Satellites also send back their data by radio.

> ### Assessment Opportunity
> You may wish to rephrase selected questions from the **As You Read** section into multiple choice or "true/false" format to use as a quiz. Use this quiz to assess student understanding and as a "motivational tool" to ensure that students complete the reading assignment and comprehend the main ideas.

> ### About the Photo
> This photo illustrates how differences in elevation can change climatic conditions over very short distances.

Investigating Climate and Weather

Investigation 4: Weather Radiosondes, Satellites, and Radar

How is the weather in the upper atmosphere measured? It is difficult and expensive to measure the weather in the upper atmosphere. But knowing what conditions are like in the upper atmosphere is important for modern weather forecasting. Since the late 1930s, meteorologists have relied mainly on radiosondes for profiles of temperature, pressure, and humidity from Earth's surface to the upper atmosphere (to altitudes of 30,000 m or 100,000 ft.). Balloons carry radiosondes up through the atmosphere. As they rise, they send back measurements by radio. Tracking of radiosonde movements from the ground shows wind speed and direction in the upper atmosphere. Eventually the balloon bursts, and the instruments fall back to Earth by parachute. (Some radiosondes are recovered and reused.) Their fall is not dangerous to humans, because the instruments are very small and light. Radiosondes are launched at the same time every 12 hours at hundreds of weather stations around the world. A similar instrument, called a dropwindsonde, is released from an aircraft to determine atmospheric conditions in areas where radiosonde data is absent (e.g., in a hurricane over the ocean).

Weather Radar

Radar (**RA**dio **D**etection **A**nd **R**anging) has become an essential tool for observing and predicting weather. Radar was invented and developed in Britain and the United States at the beginning of World War II. It was used to detect the approach of enemy airplanes. An antenna

Teacher Commentary

About the Photo

In the photo on page C45 of the student text, a technician is about to launch a weather balloon. The white object below the balloon is the instrument package.

INVESTIGATING CLIMATE AND WEATHER

sends out pulses of microwave energy. These waves are reflected from solid or liquid precipitation particles in the air and received back by the antenna. The radar equipment shows the position and distance of the particles. The results (the radar echoes) are shown as blotches on a screen, similar to a television or computer monitor. Radar echoes are electronically superimposed on a map of the area to show the location of areas of precipitation. The strength of the echoes is used to find the intensity of precipitation and to tell frozen forms (e.g., hail) from unfrozen forms of precipitation (e.g., rain). Meteorologists use radar to track the movement and follow the development of storm systems, especially small-scale systems like thunderstorms.

Weather Satellites

The first weather satellite was launched into orbit on April 1, 1960, when the United States orbited TIROS-I. Today, satellites are a routine and valuable tool in monitoring the Earth system. Satellite sensors obtain images of the Earth's weather from space. They are especially good at showing cloud cover and the temperatures of clouds and other surfaces in the sensors' field of view. A visible satellite image is like a black-and-white photograph of the planet and is available only for the areas of the planet that are in sunlight. An infrared satellite image shows the temperature of surfaces based on the invisible infrared (heat) radiation emitted by objects. Infrared satellite imagery is available both day and night and is the usual satellite picture shown on televised weathercasts. The most useful satellites are ones in a geostationary orbit. That is an orbit adjusted so that the speed and direction of the satellite matches the Earth's rotation. Then the satellite is always above the same point on Earth's surface and views the same area of the planet.

Teacher Commentary

NOTES

Investigation 4: Weather Radiosondes, Satellites, and Radar

Review and Reflect

Review

1. How do meteorologists make observations of the weather at high altitudes without going there in airplanes?
2. How are radiosondes and weather satellites similar? How are they different?
3. How is radar used to track thunderstorms?

Reflect

4. What advantages are there in looking at the world from space to observe weather?
5. In what ways, other than those you investigated, can meteorologists obtain weather data from the upper atmosphere?

Thinking about the Earth System

6. How might remote sensing by satellite be used to monitor the geosphere and hydrosphere?
7. How might satellites be used to observe features of the landscape like the distribution of vegetation and ice and snow cover?

Thinking about Scientific Inquiry

8. Give three examples of inverse relationships related to weather that you have discovered in your investigations.
9. How did you use evidence to develop ideas in this investigation?

Teacher Commentary

Review and Reflect

Review
Allow your students ample time to pull all their evidence together and arrive at conclusions and explanations. Help them make all the connections they can, based on their data.

1. Meteorologists make observations of weather at high altitudes using radiosondes and satellites.

2. Both radiosondes and satellites can be used to study the upper atmosphere. Although each records different kinds of data and information, both can give information about temperature. Satellites are expensive pieces of equipment that remain in orbit, recording images of the Earth's surface. Radiosondes, however, are small instruments that fall back to the Earth after one use; they transmit only numerical data.

3. Meteorologists can combine radar images taken over a period of time to track the movement and development of storm systems.

Reflect
Give your students time to reflect on the nature of the evidence they have generated from their investigations. Again, help them see that evidence is crucial in scientific inquiry.

4. Answers will vary. A large area of the Earth's surface can be viewed all at once, and the patterns of storm systems, as revealed by cloud cover, can be seen directly in photographs reconstructed from the data streams sent back to Earth from weather satellites.

5. Meteorologists can obtain weather data from the upper atmosphere in various ways: measurements made by high-flying aircraft, and visual observations of clouds (kind, speed of movement, direction of movement).

Thinking about the Earth System
Students should be able to make strong connections between weather and the hydrosphere, geosphere, biosphere, and atmosphere. This should help them in their growing awareness of the Earth as a set of closely linked systems. Ensure that students add their ideas to their *Earth System Connection* sheets.

6. Answers will vary. A sample response might state that images obtained from satellites could show changes in the shape and size of features like lakes and rivers.

7. Answers will vary. Sample responses include that satellite images of the same location taken over a long period of time (months) can show changes in the types and distribution of trees and grasses. Or, satellite images might be used to show the movement of glaciers or to monitor how much snow has fallen or melted in an area.

Thinking about Scientific Inquiry

As your students consider the inquiry processes they have used in **Investigation 4**, ask them to share all the ways in which they communicated their findings to each other. Ask them if some of the ways that they used were more effective than others were. Why was that? How could the class, as a whole, improve its communication strategies?

8. Air pressure and altitude; air density and altitude; air temperature and altitude (below about 12,000 m).

9. Students used evidence by examining and plotting radiosonde data to determine the relationship between temperature and altitude, by comparing satellite images and weather maps, and by comparing satellite images and radar images.

> **Assessment Tool**
> Review and Reflect Journal–Entry Evaluation Sheet
> Use the general criteria on this evaluation sheet for assessing content and thoroughness of student work. Adapt and modify the sheet to meet your needs. Consider involving students in selecting and modifying the criteria for evaluating their reflections on **Investigation 4**.

Teacher Commentary

NOTES

Teacher Review

Use this section to reflect on and review the investigation. Keep in mind that your notes here are likely to be especially helpful when you teach this investigation again. Questions listed here are examples only.

Student Achievement

What evidence do you have that all students have met the science content objectives?
_____

Are there any students who need more help in reaching these objectives? If so, how can you provide this? _____
_____

What evidence do you have that all students have demonstrated their understanding of the inquiry processes? _____
_____

Which of these inquiry objectives do your students need to improve upon in future investigations? _____
_____

What evidence do the journal entries contain about what your students learned from this investigation? _____
_____

Planning

How well did this investigation fit into your class time? _____
_____

What changes can you make to improve your planning next time? _____
_____

Guiding and Facilitating Learning

How well did you focus and support inquiry while interacting with students?
_____

What changes can you make to improve classroom management for the next investigation or the next time you teach this investigation? _____
_____

Teacher Commentary

How successful were you in encouraging all students to participate fully in science learning? _____

_____

How did you encourage and model the skills values, and attitudes of scientific inquiry? _____

_____

How did you nurture collaboration among students? _____

_____

Materials and Resources

What challenges did you encounter obtaining or using materials and/or resources needed for the activity? _____

_____

What changes can you make to better obtain and better manage materials and resources next time? _____

_____

Student Evaluation

Describe how you evaluated student progress. What worked well? What needs to be improved? _____

_____

How will you adapt your evaluation methods for next time? _____

_____

Describe how you guided students in self-assessment. _____

_____

Self Evaluation

How would you rate your teaching of this investigation? _____

_____

What advice would you give to a colleague who is planning to teach this investigation? _____

_____

NOTES

Teacher Commentary

INVESTIGATION 5: THE CAUSES OF WEATHER

Background Information

The Water Cycle
Water, in the form of liquid, solid, or vapor, is in a continuous state of change and movement. Water resides in many different kinds of places, and it takes many different kinds of pathways in its movement. The combination of all of these different movements is called the water cycle, or the hydrologic cycle.

The water cycle is called a cycle because the Earth's surface water forms a closed system. In a closed system, material moves from place to place within the system but is not gained or lost from the system. Actually, the Earth's surface water is not exactly a closed system, because relatively small amounts are gained or lost from the system. Some water is buried with sediments and becomes locked away deep in the Earth for geologically long times. Volcanoes release water vapor contained in the molten rock that feeds them. Nonetheless, these gains and losses are very small compared to the volume of water in the Earth's surface water cycle.

Evaporation (change of water from liquid to vapor) and precipitation (change of water from vapor to liquid or solid) are the major processes in the water cycle. The balance between evaporation and precipitation varies from place to place and from time to time. It's known, however, that there is more evaporation than precipitation over the surface of the Earth's oceans, and there is more precipitation than evaporation over the surface of the Earth's continents. That fact has a very important implication: there is net movement of water vapor from the oceans to the continents, and net movement of liquid (and solid) water from the continents to the oceans.

Elements of the Water Cycle
The oceans cover about three-quarters of the Earth. Ocean water is constantly evaporating into the atmosphere. If enough water vapor is present in the air, and if the air is cooled sufficiently, the water vapor condenses to form tiny droplets of liquid water. If these droplets are close to the ground, they form fog. If they form at higher altitudes due to rising air currents, they form clouds.

All of the solid or liquid water that falls to Earth from clouds is called precipitation. Snow, sleet, and hail are solid forms of precipitation. Rain and drizzle are liquid forms of precipitation.

When rain falls on the Earth's surface, or snow melts, several things can happen to the water. Some evaporates back into the atmosphere. Some flows downhill on the surface, under the pull of gravity, and collects in streams and rivers. This flowing water is called surface runoff. Most rivers empty their water into the oceans. Some rivers, however, end in closed basins on land. Death Valley and the Great Salt Lake are examples of such closed basins.

Some precipitation soaks into the ground rather than evaporating or running off. The water moves slowly downward, percolating through the spaces of porous soil and rock material. Eventually the water reaches a zone where all of the pore spaces are filled with water. This water is called ground water. Some water, called soil moisture, remains behind in the surface layer of soil.

Some of the water that soaks into the soil is absorbed by the roots of plants. This water travels upward through the stem and branches of the plant into the leaves and is released into the atmosphere in a process called transpiration.

It's been estimated that each year about 36,000 km^3 of water flows from the surface of the continents into the oceans. That represents the excess of precipitation over evaporation on the continents. This water carries sediment particles and dissolved minerals into the ocean. The sediment particles come to rest on the ocean bottom. When seawater evaporates, it leaves the dissolved materials behind. Over geologic time, this process has gradually made the oceans as salty as they are now.

Earth System Science and the Water Cycle
In Earth System science, the water cycle is viewed as a flow of matter and energy. Each place that water is held is called a reservoir. The rate at which water flows from one reservoir to another in a given time is called flux. Energy is required to make water flow from one reservoir to another. On average, the total amount of water in all reservoirs combined is nearly constant; however, the various reservoirs involved in the water cycle do not have a constant amount of water. For example, in many areas there may be more groundwater during the spring (when precipitation is high, and water use and evaporation is low) than in the summer (when precipitation is low, and evaporation and water use are high).

Clouds
Most clouds develop as a consequence of expansional cooling when air rises in the atmosphere. As a gas or a mixture of gases (e.g., air) expands, it does work on its surroundings, and the energy for that work is drawn from the store of internal energy (i.e., its heat energy). Air that ascends in the atmosphere expands and cools in response to the steady decrease in air pressure with increasing altitude.

Water vapor is a variable component of air. Any sample of air has an actual concentration of water vapor and an upper limit to that concentration (when saturated with water vapor). The saturation concentration depends on temperature; i.e., the saturation concentration increases with temperature. Cooling decreases the saturation concentration; if no water vapor is added to or removed from the air, clouds are increasingly likely to form as air cools. For this reason, clouds tend to develop in rising currents of air.

In addition to saturated conditions, another requirement for cloud formation is the presence of nuclei—tiny solid and liquid particles suspended in the atmosphere that provide surfaces upon which water vapor condenses. For a cloud droplet to form in perfectly clean air, a sufficient number of water molecules must come together at the same time to form a liquid water structure. These molecules must remain intact for long enough to grow by capturing additional water molecules before the inherent thermal agitation in the growing droplet sends all of the molecules flying apart again. For that to happen, the concentration of water molecules in the air must be extremely high. The presence of condensation nuclei gives the water molecules a stable surface upon which to condense.

More Information... on the Web
Go to the *Investigating Earth Systems* web site www.agiweb.org/ies for links to a variety of web sites that will help you deepen your understanding of content and prepare you to teach this investigation.

Teacher Commentary

Investigation Overview
Students visit a series of stations where they use models to understand the factors that influence weather. The stations investigate the effects of wind, the formation of clouds, and the effect of temperature on air pressure. **Digging Deeper** introduces the water cycle and explains how the cycle is tied to weather. The reading also reviews evaporation and condensation, and describes how clouds are related to precipitation.

Goals and Objectives
As a result of **Investigation 5**, students will develop a better understanding of the factors that cause weather.

Science Content Objectives
Students will collect evidence that:
1. The water cycle is the system of movement of water along a variety of pathways on the Earth's surface, in the Earth's oceans, and in the Earth's atmosphere.
2. Energy and water interact in the water cycle.
3. Evaporation is the process by which a substance passes from the liquid state to the vapor state.
4. Condensation—the opposite of evaporation—is the process by which a substance passes from the vapor state to the liquid state.
5. The atmosphere exerts pressure on surfaces.

Inquiry Process Skills
Students will:
1. Ask questions about the factors that influence weather.
2. Make predictions about the questions.
3. Use models to answer inquiry questions.
4. Collect data from models.
5. Analyze data from models.
6. Arrive at conclusions based on data.
7. Share findings and conclusions with others.

Connections to Standards and Benchmarks
In **Investigation 5**, students explore the factors that cause weather. These observations will start them on the road to understanding the National Science Education Standards and AAAS Benchmarks shown below.

NSES Links
- Energy is a property of many substances and is associated with heat, light, electricity, mechanical motion, sound, nuclei, and the nature of a chemical. Energy is transferred in many ways.

- Global patterns of atmospheric movement influence local weather. Oceans have a major effect on climate, because water in the oceans holds a large amount of heat.

AAAS Links

- Heat energy carried by ocean currents has a strong influence on climate around the world.

- Because the Earth turns daily on an axis that is tilted relative to the plane of the Earth's yearly orbit around the Sun, sunlight falls more intensely on different parts of the Earth during the year. The difference in heating of the Earth's surface produces the planet's seasons and weather patterns.

- Models are often used to think about processes that happen too slowly, too quickly, or on too small a scale to be observed directly, or that are too vast to be changed deliberately.

Teacher Commentary

Preparation and Materials Needed

Preparation

Investigation 5 requires the students to cycle through three different stations. You will not need to collect duplicate sets of materials for each group, as is usually the case; rather, you will just need one set of materials for each station. However, depending on the size of your class and availability of materials, you may want to set up duplicate stations so that students are kept busy throughout the class period. You can either have your students set up the stations as they come to them or get the stations ready for them before class.

At **Station 1**, students will investigate the effects of wind on the temperature recorded by a thermometer. To prepare for this station, gather the materials and tape two alcohol thermometers to a piece of cardboard.

You will need to prepare ice cubes or purchase a bag of ice for use at **Station 2** and **Station 3**.

Set up the stations so that there is plenty of room for students to circulate, and also so that you can monitor student activity at all times. Much will depend upon the space available; try to utilize the available space in as efficient a way as possible. Asking students for their ideas may allow some good ideas to surface.

It is important that you try each of the stations yourself ahead of time, so that you can provide your students with tips on how to work through the stations successfully.

Materials
- paper towels

Part A, Station 1
- water supply
- battery-powered fan
- two alcohol thermometers
- tape
- cotton batting
- piece of stiff cardboard

Part A, Station 2
- Styrofoam® picnic cooler
- brick or other heavy mass
- metal container, with lid, small enough to fit in cooler but large enough to contain the brick
- ice
- flashlight

Part A, Station 3:
- two large round balloons (same size)
- meter stick or measuring tape (or string and meter stick)
- two thermometers
- ice bath

Part B:
no additional materials are required

Teacher Commentary

NOTES

INVESTIGATING CLIMATE AND WEATHER

Investigation 5:
The Causes of Weather

 Key Question
Before you begin, first think about this key question.

What causes weather?

Think about what you know about weather patterns and weather reports from previous investigations. How does weather originate?

Share your thinking with others in your class. Keep a record of the discussion in your journal.

Materials Needed

For each station in this investigation you will need:

- paper towels
- student journal

Investigate

Part A: Visiting the Stations

1. There will be a series of stations for you to visit. The investigations at the stations will allow you to ask and answer questions about:

Teacher Commentary

Key Question

Instruct students to respond to the **Key Question** in their journals. Allow a few minutes of writing time. Have them share their ideas with a neighbor, then with the rest of their group, and finally, with the entire class. Make a list of ideas on the blackboard. Accept student ideas uncritically, even if they appear undeveloped (or are simply not correct). Neither should wrong ideas be praised. The point of this exercise, as it is with all the **Key Questions**, is to provoke thought and prepare for the investigation.

Student Conceptions about the Causes of Weather

Students may now be more familiar with weather-related factors, but they are unlikely to know what causes weather. Help students to consider the questions posed in this introduction to **Investigation 5**. Encourage them to offer any explanations they may have, and to record them for later comparison.

Answer for the Teacher Only

This is a profound question. The short answer is: the Sun. Solar energy received by the Earth is absorbed mainly at the ground surface. Higher in the atmosphere, there is a net loss of heat to outer space by long-wave radiation. That means that there must be a net transport of heat upward in the atmosphere all the time. This upward heat transport is the basic cause of vertical motions in the atmosphere. Also, over the course of the year, the low-latitude regions of the Earth gain more heat from the Sun than they lose to space, whereas the high-latitude regions lose more heat than they receive. This heat differential is the basic cause of winds and wind systems.

Assessment Tool
Key–Question Evaluation Sheet
Use this evaluation sheet to help students understand and internalize basic expectations for the warm-up activity.

Investigate

Teaching Suggestions and Sample Answers

Teaching Tip

Part A of **Investigation 5** has students using models to conduct scientific investigations. It might be useful again to discuss the scientific idea of modeling real-world processes that are too difficult to observe directly, or on a scale too large to be changed deliberately, or too dangerous. Students should begin to appreciate how useful models are as a means of observing and testing in scientific inquiry. Help students see that models must represent as realistically as possible or feasible what is being replicated.

Teaching Tip

Many of the concepts involved in **Investigation 5** cannot be discovered through hands-on activities. For this reason, students are asked to find ways of interpreting information through parallels and analogies, many of which are mathematical. It is important that they realize that this form of investigating is also part of scientific inquiry.

1. Discuss equipment and supplies and the location of each station around the room before the students begin.

Teaching Tips

You will need to allocate the stations to student groups. It is important that all stations be constructed and used. If necessary, divide your students into six new groups for this task.

You will also need to set up a schedule so that student groups can circulate through the stations in a systematic way. It may be useful to post this schedule so students can keep track of it themselves.

Students may take different lengths of time to progress through each station. Therefore, there may be some delays as students finish at one station and are waiting to "rotate." To encourage students to use this time productively, provide articles for them to read on the factors that cause weather, or questions for them to investigate using Internet resources, CD-ROM encyclopedias, or other classroom resources.

Teacher Commentary

NOTES

Investigation 5: The Causes of Weather

Station 1: The Effects of the Wind

Station 2: Cloud Formation

Station 3: Temperature and Air Pressure

Keep a record of what you do and discover at each of the stations. At the end of your "station journey," you will be looking for common threads that seem to be a part of all aspects of weather.

Station 1: The Effects of the Wind

1. Wet the back of one of your hands with room-temperature water. Leave the other hand dry.

2. Have someone turn on the fan and direct the wind toward the backs of both your hands at the same time.

   a) Observe and record any differences in how your hands feel.

   b) Explain any differences you feel.

   c) How does this experience help to explain why winds help to cool you down on a hot day? Why a cold day feels even colder when the wind is blowing?

Materials Needed

For this station your group will need:

- water supply
- battery-powered fan
- alcohol thermometers
- tape
- cotton batting

Teacher Commentary

Station 1: The Effects of the Wind

1. Students should moisten the backs of one of their hands.

2. Make sure that students hold both hands equidistant from the fan.

   a) Students should observe that the breeze on the hand that is wet feels colder than the breeze on the dry hand.

   b) Answers will vary.

   c) Moving air speeds up evaporation. Heat is removed as the water evaporates, resulting in a "wind chill" factor.

INVESTIGATING CLIMATE AND WEATHER

Inquiry
Models in Scientific Inquiry

In this investigation you are using models. A model is the approximate representation (or simulation) of a real system. For example, a weather map is a graphical model of the state of the atmosphere over a given area. Models are used by scientists to study processes that happen too slow, too quick, or on too small a scale to be observed; that are too vast to be changed deliberately; or that might be dangerous. Models are also used to organize your thinking on some complex process. These are called conceptual models.

Be careful that the fan does not touch the thermometers.

3. Using your results, develop a hypothesis related to the rest of this investigation.
   a) What do you predict will happen? Why?
4. Two thermometers are taped to a cardboard stand about 50 cm apart.
5. Read the temperatures on both thermometers
   a) Record these temperatures.
6. Hold a fan about 10 cm away from the bulb of one thermometer and turn on the blades. Observe what happens to both thermometers.
   a) Record your observations.

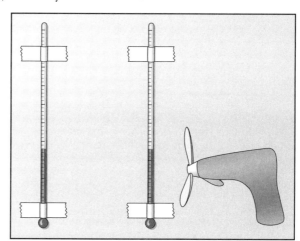

7. Next, dampen a small amount of cotton batting and tape it to one of the thermometer bulbs, so that it is exposed to the air.

   Direct a fan at the bulb of both the wrapped and the unwrapped thermometers.

   Observe what happens to the temperature readings on both thermometers.

   a) Record your observations.
   b) Compare your observations when the thermometer bulb was not dampened and when the bulb was dampened.
   c) Review your hypothesis. Explain why you think there was a difference in the temperatures of the two thermometers.

Teacher Commentary

Teaching Tip
Be sure that students read the note titled Models in Scientific Inquiry in the margin of page C50. Students need to understand how models can provide evidence in situations where observing actual phenomena is difficult. You may want to read this over with your students and engage in a brief discussion about the implications.

3. Have your students read through the rest of the instructions for **Station 1** and hypothesize on what they think will happen.

4. It would be better to tape the thermometers horizontally to a tabletop. Then there's no possibility of them slipping out of the tape and falling to the floor.

5. Temperatures will vary, but readings should be the same for both thermometers (room temperature).

6. Students should direct the fan toward the bulb for a few minutes.
    a) The temperature recorded by both thermometers should remain the same. The air temperature is the same whether the air around the bulb of the thermometer is still or moving.

7. The damp cotton should be covering the thermometer bulb and should be exposed to the air so that moisture may evaporate from it.
    a) The temperature recorded by the dry-bulb thermometer should remain the same, but the temperature recorded by the wet-bulb thermometer should decrease.
    b) See **Question 7(a)**.
    c) The moving air caused evaporation of the water in the cotton batting. Evaporation results in cooling, because the more energetic water molecules tend to be preferentially removed from the liquid, leaving behind the less energetic molecules. Keep in mind that temperature is a reflection of the average energy of motion of the constituent molecules. The cooling of the batting caused heat to be conducted from the warmer bulb to the colder batting.

Investigating Climate and Weather

Investigation 5: The Causes of Weather

Station 2: Cloud Formation

1. Place a metal container in a cooler.

   Put a brick in the bottom of the metal container to weigh it down.

   Place the lid on the metal container.

   Pack ice around the metal container in the cooler.

   Put the lid on the cooler.

2. After fifteen minutes or so the air in the metal container has been chilled. Carefully remove the lid of the cooler.

   Slowly raise the lid of the metal container a few centimeters above its rim.

   Shine a beam from a flashlight into the metal container.

   Take a deep breath; hold it for a few seconds.

   Put your mouth close to the top of the container. Very slowly and gently breathe some air out through your mouth down into the container.

   a) Record your observations.

   b) Describe how what you observed relates to the formation of clouds.

   c) Explain how you think clouds are formed.

Materials Needed

For this station your group will need:

- Styrofoam® picnic cooler
- brick or other heavy mass
- metal container, with lid, small enough to fit in the cooler but large enough to contain the brick
- bag of ice cubes
- flashlight

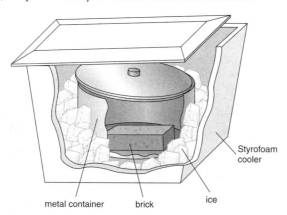

Styrofoam cooler

metal container   brick   ice

 Clean up spills immediately.

3. Using the flashlight beam, try especially to observe the size of the particles in the cloud.

   a) Record your observations.

Teacher Commentary

Assessment Tools

Journal–Entry Evaluation Sheet
Use this sheet as a general guideline for assessing student journals, adapting it to your classroom if desired.

Journal–Entry Checklist
Use this checklist as a guide for quickly checking the quality and completeness of journal entries.

Station 2: Cloud Formation

1. Either you or your students can set this station up using the instructions and text on page C51 of the student edition as a guide. For the experiment to be successful, the metal container should sit in the cooler for at least 15 minutes. Also, if the experiment is running for the entire class period, check periodically and add more ice as needed.

2. a) Students should be able to see a "cloud."

   b) Clouds form when air is saturated with water vapor and some of the water vapor condenses to liquid water. The students' breath is warm and moist, so when it is cooled by the cold air in the pot condensation is triggered. This is not unlike the cooling of a rising humid air mass, which can lead to cloud formation.

   c) Answers will vary. When air is saturated with water vapor, the water vapor condenses to form clouds. As mentioned in the background section, this process is greatly facilitated by the presence of condensation nuclei, which in this case may have been present in the students' breath.

3. a) The water droplets are of a size that's just barely large enough to be seen with the unaided eye. It helps to look at the illuminated droplets against a dark background.

INVESTIGATING CLIMATE AND WEATHER

Station 3: Temperature and Air Pressure

Materials Needed

For this station your group will need:

- two large round balloons (same size)
- metric measuring tape (or string and meter stick)
- two thermometers
- ice bath

1. Blow up two balloons to the same size and tie them shut.
   a) What do you think will happen when one balloon is cooled and the other is not? Why?
2. Measure the circumference (distance around the center) of both balloons.
   a) Record your measurements.
3. Read the temperature of two thermometers. They should both be at the same room temperature.
   a) Record this temperature.
4. Put one balloon and one thermometer into an ice bath for 10 min. Let the other balloon and thermometer stay at room temperature.

5. After 10 min, read the temperature of the thermometer in the ice bath and the temperature of the room.
   Take the balloon and thermometer out of the ice bath.
   Immediately measure the circumference of each balloon.
   a) Record the temperatures and the circumferences in your journal.
   b) What happened to the balloon in the ice bath?
   c) Did any air escape from the balloon in the ice bath?
   d) Did the density of the balloon increase or decrease because of cooling?
   e) What do you think was happening to the air inside the balloon?

 Clean up spills immediately. Dry the thermometer so it is not slippery.

Investigating Earth Systems

Teacher Commentary

Station 3: Temperature and Air Pressure

1. a) Because the density of air increases with decreasing temperature and the amount of air in the balloon is constant, the cooled balloon will contract.

2. Students can use a tape measure to measure the circumference of each balloon. Or, they can use a piece of string and a meter stick.

   a) Answers will vary. Make sure that your students measure the circumference of both balloons.

3. a) Both thermometers should start at the same temperature (room temperature).

4. Use the illustration on page C52 of the student text to help with setting up the ice bath.

5. a) Answers will vary. Students should find that the balloon in the ice bath has a smaller circumference than the balloon kept at room temperature. The temperature in the ice bath should be lower than room temperature. If the ice bath is sufficiently large, the temperature of the water in the ice bath should be at almost the freezing point.

   b) The balloon in the ice bath should contract.

   c) No, air did not escape (or it shouldn't have!)

   d) The rubber material of the balloon itself increases slightly in density, because almost all solid materials contract when cooled, so the some volume of rubber occupied a slightly smaller volume, thus causing the density (mass per unit volume) to increase slightly. The intent of the question was to ask about the density of the air inside the balloon. Because the balloon shrank, the density of the air increased—the same mass of air occupied a smaller volume.

   e) Because of the lower temperature, the motion of the air molecules in the balloon decreased. This lessened motion enables the air to occupy a smaller volume. As a result, the force of the impacts of air molecules with the inner surface of the balloon decreased, so the pressure of the air on the inner surface of the balloon was less. The walls of the balloon were then able to contract in response to the lower pressure.

Investigation 5: The Causes of Weather

6. Let both balloons stay at room temperature for five minutes.

Observe both balloons over that time.

Measure the circumference of each balloon again after five minutes.

a) Record your observations and your measurements.

b) What can this investigation tell you about warm and cold air masses? Which air mass would you expect to be denser: warm or cool? Which type of air mass might take up more space: warm or cool?

c) If a warm and a cool air mass were to meet, what do you think would happen, and why?

d) How is air pressure involved?

Part B: Sharing and Discussing Your Findings

1. When your group has completed all of the stations, share your findings with one other group in the class.

   Help each other to answer questions you might have about the science behind the weather. Look for areas that are common across the stations.

   a) In your journal record anything new that you discover during your discussion.

2. When you have finished, hold a class discussion about the science behind the weather.

   Make one large list of the common science elements for the whole class.

   a) Keep a record of the list in your student journal.

Inquiry

Sharing Findings

An important part of a scientific experiment is sharing the results with others. Scientists do this whenever they think that they have discovered scientifically interesting and important information that other scientists might want to know about. This is called disseminating research findings. In this investigation you are sharing your findings with other groups.

Teacher Commentary

6. a) After five minutes, both balloons should again have the same circumference, because the temperature of the cold balloon would have risen to room temperature again.

   b) The students might reasonably conclude that a warmer air mass would be less dense than a colder air mass. This is true, but only provided that the total mass of air in a vertical column through the two air masses is the same. That may or may not be the case in nature. The phrase "take up more space" in the final question is somewhat misleading. The intent was to query the students about which air mass would be deeper; i.e., which air mass would extend to a greater altitude. If the total mass of air in a vertical column through the two air masses is the same, the warmer air mass would extend to a greater altitude.

   c) Because the warm air mass is less dense than the cold air mass, the cold air mass will underride (i.e., move in beneath) the warm air mass, and the warm air mass will override (i.e., move up over) the cold air mass. (See **Digging Deeper, Investigation 3: Air Masses and Fronts**, pages C33 to C35 of the student text, for more details.)

   d) This is an overly vague question. The intent was to query the students about how air pressure is involved in the changes in the air density of the balloons. See the comments about **Question 5e** above. As the colder balloon gradually warms up to room temperature, the average speed of the air molecules increases. The pressure on the walls of the balloon increases until the air pressure inside the balloon is the same as that in the other balloon, which remained at room temperature.

Part B: Sharing Findings

1. a) Students' entries will vary.

2. a) The list is likely to be long and varied. At a minimum, it should include material that has to do with such common science elements as the following:
   - solar heating of the Earth, and its consequences for winds
   - the vertical distribution of temperature, pressure, and density of the atmosphere
   - the various types of air masses and their movements
   - the nature of high-pressure and low-pressure areas

Teaching Tip

In this investigation, students are responsible for applying inquiry processes. By now, they should have had enough experience in using these from earlier investigations. This is their chance to apply them to their own work. Some students may be so focused on the practical nature of their investigation that they may forget about this inquiry aspect. Your role is to give them constant reminders about "how" they are investigating, and to encourage them to think out the processes they are using.

Assessment Tool

Investigation Journal–Entry Evaluation Sheet

Use this sheet to help students learn the basic expectations for journal entries that feature the write-up of investigations. It provides a variety of criteria that both you and your students can use to ensure that their work meets the highest possible standards and expectations. Adapt this sheet so that it is appropriate for your classroom, or modify the sheet to suit a particular investigation.

Teacher Commentary

NOTES

INVESTIGATING CLIMATE AND WEATHER

Digging Deeper

THE WATER CYCLE
The Main Loop of the Water Cycle

As You Read...
Think about:
1. What is the water cycle?
2. What is the difference between evaporation and condensation?
3. On the Earth's surface, where would you likely find more evaporation than condensation occurring? Where would you likely find more condensation than evaporation?
4. How are clouds formed?
5. How is rain formed?
6. How does temperature affect air pressure?

A "closed system" consists of a container that allows energy, but not matter, to pass back and forth across the walls of the container. The Earth's atmosphere, ocean, and land surface act as an almost closed system. Water moves along a variety of pathways in this closed system. This system of movement is the global water cycle.

There is one main loop in the water cycle. Water evaporates at the Earth's surface and then moves as water vapor into the atmosphere. The water vapor condenses into clouds and falls from clouds as precipitation back to Earth's surface. Water that falls onto the continent can follow a number of pathways: some water evaporates back into the atmosphere, some is temporarily stored in lakes, reservoirs, and glaciers, some seeps into the ground as soil moisture and ground water, and some runs off into rivers and streams. Ultimately, all the water on land drains into the ocean.

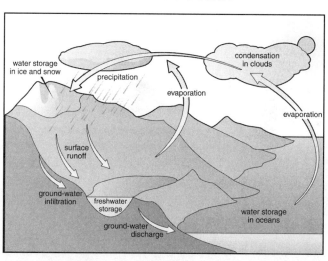

Evidence for Ideas

C 54
Investigating Earth Systems

Teacher Commentary

Digging Deeper
This section provides text, a diagram and several photos that give students greater insight into the water cycle, particularly with respect to evaporation, condensation, and precipitation. You may wish to assign the **As You Read** questions as homework to help students focus on the major ideas in the text.

As You Read...
1. The water cycle is the system of movement of water along a variety of pathways on the Earth's surface, in the Earth's oceans, and in the Earth's atmosphere.

2. Evaporation is the process by which a substance passes from the liquid state to the vapor state. Evaporation occurs if the flux of water molecules passing from the liquid to the vapor exceeds the flux of water molecules passing from the vapor to the liquid. Condensation—the opposite of evaporation—is the process by which a substance passes from the vapor state to the liquid state. Condensation occurs if the flux of water molecules passing from the vapor to the liquid exceeds the flux of water molecules passing from the liquid to the vapor.

3. Answers will vary. The most straightforward answer is that there is an excess of evaporation over condensation over the oceans and condensation surpasses evaporation over the continents. Evaporation is more likely to occur in areas with high temperatures and low air humidity. Condensation immediately at the Earth's surface is mainly in the form of dew, which forms on clear nights on the ground surface if the relative humidity is sufficiently high at the outset. Condensation is more common in rising, moist air, well above the Earth's surface.

4. Clouds form when air rises upward in the atmosphere and is thereby cooled until the saturation point is reached. Once a cooling air mass reaches the condition of saturation condensation can occur and clouds can form.

5. Raindrops form when water droplets that have condensed from water vapor in the air fall to the Earth's surface. For raindrops to be large enough to fall at an appreciable speed, however, they must grow to be much larger than the tiny water droplets that first condense to form clouds. One important way they grow is by collisions between slightly larger droplets and slightly smaller droplets. The larger the droplet, the faster the speed of fall. Larger droplets therefore catch up to, and merge with, smaller droplets as they fall.

6. Heat increases air pressure; because of the greater speed of air molecules, the collisions of the air molecules with their container are more forceful.

Assessment Opportunity

You may wish to rephrase selected questions from the **As You Read** section into multiple choice or "true/false" format to use as a quiz. Use this quiz to assess student understanding and as a "motivational tool" to ensure that students complete the reading assignment and comprehend the main ideas.

About the Photo

Use **Blackline Master** *Climate and Weather* 5.1 in this Teacher's Edition to make an overhead of the water cycle diagram on page C54 of the student text. Incorporate this overhead into a lecture reviewing the major elements of the water cycle. **Blackline Master** *Climate and Weather* 5.2 can be used as a quick assesment to this discussion.

Teacher Commentary

NOTES

Investigation 5: The Causes of Weather

Each year, there is an excess of evaporation over precipitation on the oceans, and an excess of precipitation over evaporation on the continents. The net gain of water on the continents equals the net loss of water from the ocean. Under the influence of gravity, the excess water on land flows to the sea.

Evaporation and Condensation

With the range of temperature and pressure conditions in the Earth system, water coexists in all three phases (solid, liquid, vapor). It is continually changing from one phase to another. In the solid phase, water molecules vibrate about fixed locations, so an ice cube is crystalline and retains its shape (as long as its temperature is below freezing). In the liquid phase, water molecules have considerably more energy. The molecules are free to move around one another. For this reason, water takes the shape of its container. In the vapor phase, water molecules have the most energy. Even a small amount of water vapor spreads evenly throughout the volume of any container.

At the interface between water and air (e.g., the surface of the ocean or lake), water molecules move in two directions. Some molecules leave the water surface to become vapor, and some molecules leave the vapor phase to become liquid. Evaporation occurs if the flux of water molecules becoming vapor exceeds the flux of water molecules becoming liquid. Condensation occurs if the flux of water molecules becoming liquid exceeds the flux of water molecules becoming vapor. Surface water temperature largely controls the rate of evaporation, because more energetic water molecules (in warmer water) escape a water surface more readily than less energetic water molecules (in colder water).

A similar type of two-way exchange of water molecules takes place at the interface between ice (or snow) and air. Sublimation occurs when more water molecules

Teacher Commentary

NOTES

INVESTIGATING CLIMATE AND WEATHER

become vapor than solid, and deposition occurs when more water molecules become ice than vapor.

Water vapor enters the atmosphere mostly by evaporation and sublimation of water at the Earth's surface along with transpiration of water by plants. There is an upper limit to the water-vapor component of air. This limit depends largely on temperature. Air is saturated when the water vapor component of air is at its upper limit. The saturation concentration of air increases with temperature, so that warm saturated air has more water vapor than cold saturated air. It follows that sufficient cooling of unsaturated air causes it to become saturated. When air is saturated, excess water vapor condenses (or deposits) into clouds. This happened when you breathed into the cold metal container in the cooler at Station 2.

The relative humidity is defined as the ratio of the actual amount of water vapor to the amount of water vapor at saturation. It is always expressed as a percentage. Suppose that a 1-kg sample of air at 20°C (68°F) has 7.5 g of water vapor. At that temperature a 1-kg sample of air would be saturated if it had about 15 g of water vapor. Hence, the relative humidity of the sample is (7.5 g/15 g) × 100% = 50%. When air is saturated, its relative humidity is 100%. As unsaturated air is cooled, its relative humidity increases. At a relative humidity of 100% water vapor condenses into liquid water droplets or deposits into ice crystals.

Teacher Commentary

> **About the Photo**
>
> The photo on page C56 of the student text shows cumulus clouds, seen from below. If the view had been from the side rather from below, these clouds would appear to have a flat base and puffy, irregular tops.

Investigation 5: The Causes of Weather

If you fill a glass with ice water on a warm and humid day, small drops of water soon appear on the outside surface of the glass. That water did not leak through the glass. It condensed from the air. The relatively cold surface of the glass chilled the air in contact with the glass causing its relative humidity to increase to 100%. At saturation, water vapor condensed to the small liquid drops. Dew or frost on the grass on a chilly morning forms in the same way, when the ground surface is chilled by radiating its heat out to space on a clear night.

Clouds and Precipitation

When air is heated in a rigid closed container, it tries to expand, because air molecules move faster when the air is warmer. The molecules collide more and more strongly with the walls of the container. This increases the pressure on the inside surfaces of the container. The atmosphere has no walls (except the Earth's surface) and is free to expand when heated. For this reason, at constant pressure (at the Earth's surface, for example) the density of air increases with falling temperature. The density of air decreases with rising temperature. The balloons at Station 3 are flexible, so that the air inside was free to expand and contract in response to changes in temperature.

If air is heated near the ground, it expands. It is then less dense than the air nearby. The cooler, denser air nearby

Teacher Commentary

About the Photo

The photo on page C57 of the student text shows dew collected on a spider web. On a still, clear night, the temperature often falls to the saturation temperature of the air next to the chilled surfaces. Water then condenses on the surfaces. Owing to surface tension, on many kinds of surfaces the water "balls up" rather than forming a continuous film.

INVESTIGATING CLIMATE AND WEATHER

pushes under the warmer, less dense air, causing it to rise in the atmosphere. Earlier in this module you learned that the air pressure decreases upward in the atmosphere. As the air rises, it expands in response to the falling pressure. As the air expands, it cools. The reason for the cooling is that the air uses some of its thermal energy doing the work of pushing on the surrounding air. That leaves the air with less thermal energy; in other words, its temperature falls.

Most clouds consist of water droplets that have condensed from water vapor in the air. The droplets fall very slowly toward the Earth. Larger droplets fall faster than smaller droplets. When a larger droplet catches up with a smaller droplet on the way down, the two combine to form an even larger droplet. That one then falls even faster, and sweeps up even more small droplets. Soon a large drop is formed, which falls to Earth' surface as rain.

Raindrops that fall through very cold air near the Earth's surface can freeze to form little grains of ice called ice pellets or sleet. Snowflakes, however, are not frozen raindrops. Snowflakes grow in clouds by addition of water molecules onto their crystal surfaces, directly from the water vapor of the surrounding air—a process called deposition.

Teacher Commentary

About the Photo

Freezing rain coats the surfaces of trees and roads with a layer of ice. The added weight of the ice often brings down tree branches and powerlines. Freezing rain poses a variety of serious hazards for people, particularly drivers.

Making Connections...*with a Real Life Situation*

Engage students' interest in precipitation (described in the **Digging Deeper** on page C58) by asking them to share their experiences with severe weather that involved a lot of rain, snow, sleet, or hail. Questions for discussion include: Can you recall the day when you saw the most rain, snow, or hail? What did you notice about the sky during this event? Give an example of how the precipitation that fell during the storm affected you, either before the storm, or afterward.

Investigation 5: The Causes of Weather

Review and Reflect

Review

1. Describe briefly what you discovered at each station.

Reflect

2. Explain why the global water cycle is called a cycle. You may wish to use a diagram to illustrate your answer.

3. What is the role of heat energy in weather? Give some examples.

4. Make a list of all of the different kinds of pathways you can think of that a water molecule might follow as it goes through the global water cycle. Remember that the water molecule can exist as water vapor, liquid water, or ice, and that it can occur at or near the Earth's surface.

Thinking about the Earth System

5. What role do plants play in the global water cycle?

6. What controls the part of precipitation that runs off into river and stream channels versus the part that infiltrates into the ground?

7. In what sense does the global water cycle involve the flow of energy as well as water?

Thinking about Scientific Inquiry

8. How did you use modeling in this investigation?

9. Why do you think sharing findings is an important process in scientific inquiry?

Teacher Commentary

Review and Reflect

Review
Allow your students ample time to pull all their evidence together and arrive at conclusions and explanations. Help them to make all the connections based upon their data.

1. At **Station 1**, students learned that moving air (wind) lowers the temperature of a moistened material. At **Station 2**, students modeled how clouds form. At **Station 3**, students learned that temperature affects air pressure.

Reflect
Give your students time to reflect on the nature of the evidence they have collected from their investigations. Help them to see that evidence is crucial in scientific inquiry.

2. The global water cycle is called a cycle because water continually moves along a number of different pathways that form "loops." The water cycle is essentially a closed system, in the sense that the water moves from place to place along the many pathways but the total volume of water remains virtually constant.

3. Answers will vary. A few possible responses follow:
   - Heat energy is needed to convert liquid water into water vapor (evaporation).
   - When water vapor loses heat energy, it becomes liquid water again (condensation).

4. Below is a list of some of more important pathways in the water cycle. There are many others as well.
   - evaporation at the ocean surface or the land surface
   - transport as vapor by winds in the atmosphere
   - transport as water droplets in clouds
   - precipitation as rain or snow on the oceans or on the continents
   - movement as ice in glaciers
   - melting and runoff from glaciers
   - runoff as liquid water in streams
   - infiltration of rainwater into the ground
   - movement as groundwater

Thinking about the Earth System
Investigation 5 will have enhanced your students' knowledge of the origins of weather. They now need to reflect on this in terms of the Earth system. Help them connect what they have learned with the geosphere, the atmosphere (water cycle), the hydrosphere (water on land), and the biosphere (living things that are connected with water, land, and air).

5. Plants take up water through their roots and then release the water back to the atmosphere by transpiration.

> **Teaching Tip**
> If you would like to have your students do an activity on transpiration, here is a simple but instructive one. Water a vigorously growing potted plant until all of the excess water drains out through the holes in the bottom of the pot. Cover the soil surface tightly with plastic wrap so that no water can escape by evaporation. Weigh the pot on an accurate scale. Place the pot in the Sun, so that the plant can continue with its life processes. Weigh the pot again in one or two days. Ask the students to explain the loss in weight, given that evaporation from the soil surface was prevented. What's happening is that the plant releases water vapor from tiny openings, called stomata.

6. Several factors control what part of precipitation goes into rivers, via surface runoff, and what part infiltrates into the ground to become groundwater. The most important of these are the rate of precipitation (the more intense the precipitation, the greater the proportion of surface runoff), the permeability of the ground surface (the greater the permeability, the greater the proportion of infiltration), and the slope of the ground surface (the greater the slope, the greater the proportion of surface runoff).

7. Moving water has kinetic energy (i.e., energy of motion). That energy can be harnessed by waterwheels or by turbines that turn electrical generators. Water also has potential energy (i.e., energy of position). Water at high elevations has the potential to flow downhill and release some of its energy. Latent heat is also important. It takes energy to evaporate water, and when the water condenses again, that latent heat is released.

> **Teaching Tip**
> Remind students to enter any new connections that they have found on the *Earth System Connection* sheet in their journals.

Thinking about Scientific Inquiry

Give students time to reflect on the nature and limitations of the models they used. Again, help them see that modeling is useful to scientific inquiry.

8. Experimenting with the effect of wind on dry and wet thermometers; simulating the generation of cloud; experimenting with the effect of air pressure on air density.

Teacher Commentary

9. Disseminating knowledge to other scientists, to help them in their work; to have other scientists provide a critique of your work, which might lead you to change or redirect your thinking.

> **Assessment Tool**
> **Review and Reflect Journal–Entry Evaluation Sheet**
> Use the general criteria on this evaluation sheet for assessing content and thoroughness of student work. Adapt and modify the sheet to meet your needs. Consider involving students in selecting and modifying the criteria for evaluating their reflections on **Investigation 5**.

Teacher Review

Use this section to reflect on and review the investigation. Keep in mind that your notes here are likely to be especially helpful when you teach this investigation again. Questions listed here are examples only.

Student Achievement

What evidence do you have that all students have met the science content objectives?

Are there any students who need more help in reaching these objectives? If so, how can you provide this?

What evidence do you have that all students have demonstrated their understanding of the inquiry processes?

Which of these inquiry objectives do your students need to improve upon in future investigations?

What evidence do the journal entries contain about what your students learned from this investigation?

Planning

How well did this investigation fit into your class time?

What changes can you make to improve your planning next time?

Guiding and Facilitating Learning

How well did you focus and support inquiry while interacting with students?

What changes can you make to improve classroom management for the next investigation or the next time you teach this investigation?

Teacher Commentary

How successful were you in encouraging all students to participate fully in science learning? _____

_____

How did you encourage and model the skills values, and attitudes of scientific inquiry? _____

_____

How did you nurture collaboration among students? _____

_____

Materials and Resources

What challenges did you encounter obtaining or using materials and/or resources needed for the activity? _____

_____

What changes can you make to better obtain and better manage materials and resources next time? _____

_____

Student Evaluation

Describe how you evaluated student progress. What worked well? What needs to be improved? _____

_____

How will you adapt your evaluation methods for next time? _____

_____

Describe how you guided students in self-assessment. _____

_____

Self Evaluation

How would you rate your teaching of this investigation? _____

_____

What advice would you give to a colleague who is planning to teach this investigation? _____

_____

NOTES

Teacher Commentary

INVESTIGATION 6: CLIMATES
Background Information

Elements of Climate

Climate is more than just the long-term average of temperature and precipitation in a region. Several other factors are important in determining climate as well. The range of temperature on a scale of days to weeks is also important, as is the yearly range in temperature. Frequency of precipitation is important, in addition to yearly averages: does rain fall often in small amounts, or does it tend to fall less often but in large amounts? The relative timing of yearly temperature and precipitation is very important for growth of trees and shrubs: in areas with large seasonal variations in precipitation, the wet season might coincide with the warm growing season, or the wet season might coincide with the cold season, in which trees and shrubs are dormant. Arid or semiarid regions with the former conditions can support much more woody vegetation, given the same level of annual precipitation, than regions characterized by the latter conditions.

Several other factors—like wind speed, wind direction, cloud cover (which is correlated with amount of precipitation, but in some areas not strongly), relative humidity, and frequency of occurrence of unusually strong storms like tropical cyclones (hurricanes and typhoons)—are also important factors in determining climate. The most sensitive indicator of climate is vegetation. In fact, the connection between vegetation and climate is so strong that climatologists attempting to develop classifications of climate have, to a great extent, been guided by vegetation type in choosing criteria by which to classify climates.

Specific Heat Capacity

One factor related to weather and climate is how different forms of matter that make up the Earth hold the heat from the Sun. Heat is a form of energy. It is a manifestation of the thermal energy of motion of the atoms and molecules that constitute matter. (See **Investigation 1: Background Information** on the connection between temperature and heat.) Heat is transferred between different bodies of matter by conduction or by radiation.

In conduction, heat is transferred from a hotter body to a colder body in direct contact with one another. That happens because the faster-moving atoms or molecules of the hotter body transfer some of their energy of motion to the slower-moving atoms or molecules of the colder body. Heat energy is thus transferred from the hotter body to the colder body, and in the process the colder body is made hotter and the hotter body is made colder.

Radiation causes heat transfer between bodies that are separated in space. All matter radiates energy in the form of electromagnetic waves. The part of the spectrum of electromagnetic waves that we feel as heat is the long-wave part of the spectrum called infrared. The radiating body loses heat and its temperature increases, while the body on which the radiation falls gains heat and its temperature rises.

Different kinds of matter differ in the relationship between the gain and loss of heat, on the one hand, and the increase or decrease of temperature, on the other hand, during heat transfer. In a general way, that has to do with the nature of the bonding forces among the constituent atoms and molecules of the given material.

The specific heat capacity of a material is the amount of heat needed to raise the temperature of the material by one degree Celsius. Materials with a high specific heat capacity act as good reservoirs of heat, because it takes a lot of heat to change their temperature. Liquid water has an extremely high heat capacity, higher than almost any other substance. The reason has to do with the existence of what are called hydrogen bonds between the electrically positively charged "side" of the water molecule and the negatively charged "side." The structure of ice involves such hydrogen bonds. When ice melts, some but not all of the hydrogen bonds are broken. As heat is added to water, the heat must act not just to increase the thermal motions of the molecules but also to break more hydrogen bonds.

More Information…on the Web
Go to the *Investigating Earth Systems* web site www.agiweb.org/ies for links to a variety of web sites that will help you deepen your understanding of content and prepare you to teach this investigation.

Teacher Commentary

Investigation Overview
Students look at a map that shows the various climatic regions of the world to determine what weather elements are used to define a climatic region. They design an experiment that allows them to investigate the ability of different materials to hold heat. Finally, each student group selects a climatic region and generates a set of six clues that their classmates then use to identify which climatic region they are describing. **Digging Deeper** reviews the difference between climate and weather, explores how temperature and precipitation are used to define the climate of an area, and examines how climate affects vegetation of an area.

Goals and Objectives
As a result of **Investigation 6**, students will recognize the difference between climate and weather and understand the factors that define and affect climate.

Science Content Objectives
Students will collect evidence that:
1. Climate is characterized by precipitation and air temperature.
2. Climate is affected by elevation, latitude, and proximity to water bodies and to mountain ranges.
3. Different forms of matter retain heat to different extents.

Inquiry Process Skills
Students will:
1. Use maps to investigate questions about climate.
2. Compare and contrast climatic regions around the world.
3. Experiment with the way in which different states of matter (solid, liquid, and gas) retain heat.
4. Collect data from experiments.
5. Arrive at conclusions based on data analysis.
6. Communicate findings and results to others.

Connections to Standards and Benchmarks
In **Investigation 6**, students learn what factors define and affect climate. These observations will start them on the road to understanding the National Science Education Standards and AAAS Benchmarks shown below.

NSES Link
- Global patterns of atmospheric movement influence local weather. Oceans have a major effect on climate, because water in the oceans holds a large amount of heat.

- Clouds, formed by the condensation of water vapor, affect weather and climate.

AAAS Link

- Heat energy carried by ocean currents has a strong influence on climate around the world.

- Models are often used to think about processes that happen too slowly, too quickly, or on too small a scale to be observed directly, or that are too vast to be changed deliberately.

- The cycling of water in and out of the atmosphere plays an important role in determining climatic patterns. Water evaporates from the surface of the Earth, rises and cools, condenses into rain or snow, and falls again to the surface. The water falling on land collects in rivers and lakes, soil, and porous layers of rock, and much of it flows back into the ocean.

Teacher Commentary

Preparation and Materials Needed

Preparation
In **Investigation 6**, your students will be connecting what they have learned about weather with the broader issue of climate.

Part A
No advance preparation is required for **Part A** of Investigation 6.

Part B
You will need to assemble all of the materials before class. The containers used in **Part B** can be 16-oz. deli containers—just be certain that the container is tall enough to accommodate the thermometers. You will need to punch holes in the lids of the containers prior to this investigation. A Phillips® screwdriver will work for this, but it is important for safety reasons that you, not your students, punch the holes.

Part C
Assemble resource materials about climate before class. *The Investigating Earth Systems* web site can serve as an excellent resource.

Materials
Part A:
- blank global map*
- colored pencils or markers

Part B:
- three heat-resistant containers with a pencil-sized hole punched in the center of each lid
- water supply
- three thermometers
- heat lamp
- graph paper

Part C:
- climate resources (books, CD-ROMs, Internet access, etc.)
- poster board and presentation supplies

---

* The *Investigation Earth Systems* web site provides suggestions for obtaining these resources.

Investigating Climate and Weather – Investigation 6

INVESTIGATING CLIMATE AND WEATHER

Investigation 6:
Climates

Key Question
Before you begin, first think about this key question.

What is the difference between weather and climate?

Think about what you have learned about the nature of weather and how it is reported. How does it differ from climate?

Share your thinking with others in your class. Keep a record of the discussion in your journal.

Materials Needed

For this investigation, all groups will need:

- a blank global map
- colored pencils or markers
- three heat-resistant containers with a pencil-size hole punched in the center of each lid (three per group)
- water supply
- sand
- three thermometers
- heat lamp
- graph paper
- climate resources (books, CD-ROMs, Internet access, etc.)
- poster board and presentation supplies

Investigate

Part A: Climatic Regions of the World

1. Look at the map on the following page showing the various climatic regions of the world. Read over the names of the types of climate.

    a) What weather elements do the name of the climates imply?

Teacher Commentary

Key Question

Begin by asking students to respond to the **Key Question**, "What is the difference between weather and climate?" Tell students to write down their ideas in their journals. After a few minutes, discuss students' ideas in a brief conversation. Emphasize thinking and sharing of ideas. Avoid seeking closure (i.e., the "right answer"). Record all of the ideas that students share on an overhead transparency or on the chalkboard. Have students record this information in their journals.

Student Conceptions about the Differences between Weather and Climate

Most students will now have a reasonable understanding of weather, but they may have little understanding about the relationship between weather and climate or how weather and climate are different. They may know that different parts of the world have different climates, and that climate generally becomes colder away from the Equator and toward the North Pole and the South Pole. However, they are unlikely to understand the influences of land masses and oceans upon climate.

Answer for the Teacher Only

Climate can be considered as the integrated effects of the prevailing or average weather conditions over a long period of time (many years).

Investigate

Teaching Suggestions and Sample Answers

1. The map shown on page C61 of the student text is based on a modified Köppen classification system. The climatic regions are defined below and detailed in **Blackline Master** *Climate and Weather* 6.1. You may wish to share these definitions with your students, if they seem to be having a difficult time understanding the terms used to name the different climatic regions.

   ### Modified Köppen Classification System: Climatic Regions

   Tropical Rain Forest: Over 2.4 in. of precipitation received in all months of the year. Average temperature of coolest month of the year does not fall below 64.4°F.

   Tropical Monsoon: At least one month of the year receives less than 2.4 in. of precipitation. Average temperature of coolest month of the year does not fall below 64.4°F.

   Tropical Savanna: Summer months are wet, winter months are dry. Average temperature of coolest month of the year does not fall below 64.4°F.

   Desert: Dry, evaporation is greater than precipitation by at least half.

   Steppe: Semiarid, evaporation is greater than precipitation by less than half.

   Mediterranean: Dry summer. Average temperature of coolest month of the year is greater than 32°F, but not more than 64.4°F. Average temperature of warmest month of the year is greater than 50°F.

Humid Subtropical: Some regions experience a dry winter period, while others have no dry season and precipitation exceeds 1.2 in. each month of the year. Temperature of coolest month of the year does not fall below 32°F, and does not rise above 64.4°F. Temperature of warmest month of the year is greater than 50°F.

Marine West Coast: No dry season, precipitation exceeds 1.2 in. each month of the year. Average temperature of coolest month of the year does not fall below 32°F, and does not rise above 64.4°F. Average temperature of warmest month of the year is greater than 50°F.

Humid Continental, Warm Summer: Some regions experience a dry winter period, while others have no dry season and precipitation exceeds 1.2 in. each month of the year. Average temperature of coolest month of the year does not fall below 32°F. Average temperature of warmest month of the year is greater than 50°F. Temperatures vary greatly.

Humid Continental, Cool Summer: Some regions experience a dry winter period, while others have no dry season and precipitation exceeds 1.2 in. each month of the year. Average temperature of coolest month of the year does not fall below 32°F. Average temperature of warmest month of the year is greater than 50°F. Temperatures vary greatly.

Subarctic: Some regions experience a dry winter period, while others have no dry season and precipitation exceeds 1.2 in. each month of the year. "Snow" climate. Average temperature of coolest month of the year does not fall below 32°F. Average temperature of warmest month of the year is greater than 50°F. Temperatures vary greatly.

Tundra: Cold, "ice" climate. Average temperature of the warmest month of the year is below 50°F.

Ice cap: Cold, "ice" climate. Average temperature of the warmest month of the year is below 50°F.

> **Teaching Tip**
> It is difficult to appreciate the distribution of climate throughout the world on a map as small as this. If possible, provide a larger map of world climates for students to use here. Such maps are usually included in a good world atlas. An atlas may also contain more information about climate types.

Teacher Commentary

NOTES

Investigating Climate and Weather

Investigation 6: Climates

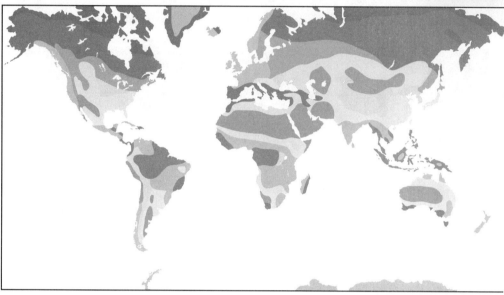

- Tropical Rain forest, Monsoon (wet)
- Tropical Savanna (wet summer, dry winter)
- Steppe (semi arid)
- Desert (dry)
- Mediterranean (dry summer)
- Humid Subtropical
- Marine West Coast (no dry season)
- Humid Continental, Warm Summer
- Humid Continental, Cool Summer
- Subarctic (snow climate)
- Tundra (cold ice climate)
- Ice Cap

b) What does this tell you about how climate is defined?

c) Write down any other questions you may have about the map, its legend, and the climatic regions of the world.

2. Find the area on the map where you live.

With your group, discuss any ideas that you might have about the weather where you live. Write down your ideas.

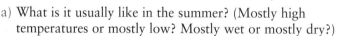

a) What is it usually like in the summer? (Mostly high temperatures or mostly low? Mostly wet or mostly dry?)

b) What is it usually like in the winter?

c) What kinds of plants and animals live in your area?

d) How do people adapt to changes in the seasons?

3. Make a class profile of your ideas about your local climate. Think carefully and record your answers to the questions on the following page.

Inquiry

Scientific Questions

Scientific inquiry starts with a question. Scientists take what they already know about a topic, then form a question to investigate further. The question and its investigation are designed to expand their understanding of the topic. You are doing the same.

Investigating Earth Systems

C 61

Teacher Commentary

2. Answers to these questions will vary depending upon the area where you live.

Teaching Tip

Students need to understand that plants and animals are adapted for the climatic conditions in the area where they live. For this reason, plants and animals are strong indicators of climate type.

INVESTIGATING CLIMATE AND WEATHER

a) What is it about your area that helps vegetation or crops to grow well, or not well?

b) What is it about your area that makes it suitable for the animals that live there?

c) What kind of clothing do people wear in winter in your area? In the summer?

d) At what latitude do you live?

e) Are you close to a large body of water (e.g., ocean, Great Lakes), or are you landlocked?

f) What is your elevation above sea level?

4. Find other parts of the world that have the same climate as your area.

a) Shade these regions in on a blank copy of the map of the world.

b) What do all of the areas have in common?

c) What other factors might affect areas that have your climate?

d) Would you expect to find the same climate as yours in the polar regions? Why or why not?

5. As you saw from your climate map, the two main weather elements that are used to describe climate are air temperature and precipitation.

a) Why do you think it is that some areas of the world are typically hot and others are typically cold?

b) Why are some areas typically dry and others typically moist?

Part B: Ability of Different Materials to Hold Heat

1. One factor related to weather and climate is how different forms of matter that make up the Earth (solids, like soil; liquids, like water; and gases, like air) hold the heat from the Sun.

Work together in your group to design an experiment to investigate what happens when these three kinds of matter are heated and left to cool.

 Be sure that your teacher checks your procedure before you begin.

Teacher Commentary

3. Answers will vary depending upon the area where you live. This series of questions is designed to help students focus on the factors that determine climate, as well as the adaptations that living things make to their climatic conditions. It is important that students understand that factors affecting climate include latitude, elevation, proximity to bodies of water, continental position (eastern or western sides), and proximity to mountain ranges.

4. a) Areas that have the same climate as where you live will vary.
   b) Answers will vary.
   c) Answers will vary.
   d) No, most likely polar regions will have climates much different than your local climate.

5. Students need to understand that temperature and precipitation are the two key weather elements that are usually used to describe climate. This section helps to prepare your students for **Part B** of **Investigation 6**—measuring varying levels of heat retention of solid, liquid, and gaseous media.
   a) Answers will vary, but students will likely note the strong correlation between latitude (or position on the Earth) and temperature.
   b) Students are less likely to understand what controls the distribution of precipitation around the globe, but they may refer to proximity to an ocean or other body of water

Assessment Tools
Journal–Entry Evaluation Sheet
Use this sheet as a general guideline for assessing student journals, adapting it to your classroom if desired.

Journal–Entry Checklist
Use this checklist as a guide for quickly checking the quality and completeness of journal entries.

Investigation 6: Climates

You will be using the following materials:
- heat-resistant containers with lids that have a thermometer-size hole in the center;
- sand, water, and air;
- thermometers;
- heat lamp.

Make sure that the experiment you design is "fair" (objective and systematic).

Complete your design. Decide on the steps you will take from start to finish. Be sure to include any safety precautions you will take.

As a group, decide what you think will happen.

a) Record your group's prediction and the reason for the prediction. This is your hypothesis.

b) Include your procedure in your journal.

c) In your journal create a table for recording your data.

2. With the approval of your teacher conduct your experiment.

a) Record all your observations.

b) Graph your data and look for any patterns that emerge.

c) Which substance had the highest beginning temperature? Why do you think that was?

d) Which substance cooled most slowly? most quickly? What ideas do you have about that?

e) How can what you just observed about land, water, and air help you to understand some of the factors that affect climate?

Inquiry

Hypotheses

When you make a prediction and give your reasons for that prediction, you are forming a hypothesis. A hypothesis is a statement of the expected outcome of an experiment or observation, along with an explanation of why this will happen.

A hypothesis is never a guess. It is based on what the scientist already knows about something. A hypothesis is used to design an experiment or observation to find out more about a scientific idea or question. Guesses can be useful in science, but they are not hypotheses.

Dependent and Independent Variables

In all experiments, there are things that can change (vary). These are called variables. In a "fair" test, scientists must decide which things will be varied in the experiment and which things must remain the same. In this investigation you will make measurements to determine how well each kind of matter holds heat. This is the dependent variable. The kind of matter that you are testing is called the independent variable. All other variables must be controlled; that is, nothing else should change.

Teacher Commentary

Part B: Ability of Different Materials to Hold Heat

1. Student experiments may vary slightly. Instructions for a "sample" experiment follow:

   Number the three containers. Fill Container #1 with sand, Container #2 with water, and Container #3 with nothing but air. Insert a thermometer through the hole in the lid of each container. Make sure that the thermometer is inserted to the same depth in each container.

   Place the containers under heat lamps for 10 minutes. At the end of this time, remove the heat lamps and record the beginning temperature in a data table. Every three minutes, take another temperature reading and record it. Do this until all three thermometers reach the same temperature (i.e., room temperature).

   a) Student predictions will vary depending on their current knowledge and how they chose to set up their experiment.

   b) Students should include detailed instructions for carrying out the experiment. Remind them that the instructions should be such that someone else could duplicate the experiment.

   c) A sample data table is shown below. Student tables will vary depending upon the experiment they design.

Temperature Change Data Table			
Temperature	Container #1 (sand)	Container #2 (water)	Container #3 (air)
0 minutes (starting)			
3 minutes			
6 minutes			
9 minutes			
12 minutes			
etc.			

Teaching Tip

Be sure that the heat lamps are not so close to the containers that they soften the plastic of the containers. Set the equipment up so that the heat lamps simulate sunlight as much as possible. If the weather is appropriate, you could use actual sunlight instead of heat lamps.

Help students appreciate that a test is fair (i.e., objective or unbiased) only if certain variables are controlled:
- all containers must be identical
- the thermometers must be placed in the same way for each
- the containers should all receive light in the same way
- students should not move the containers in any way when reading the thermometers.

> **Teaching Tip**
> When you emphasize that these variables must be controlled, students should be able to see what makes the test fair, and what might interfere with this. Help them see that a test becomes flawed if attention to detail is not observed.

2. Be sure to check and approve all designs before students begin their experiments.

    a) Emphasize the need for accurate and systematic recording of observations.

    b) Data patterns will vary depending on how students have elected to complete their experiment.

    c) By "beginning temperature" is meant the temperature immediately after the heat source is removed. Answers will vary depending on how long the students wait to take their initial temperature readings. Ideally, all three materials would be at room temperature at the start of the investigation. After several minutes under the heat lamp, the sand would have the highest temperature. That has less to do with the heat capacity of the three materials; it has much more to do with transparency. Most of the light that is incident upon the air and water passes through and out the other side, because of the transparency. On the other hand, if the initial temperatures are taken after a much longer period of time, long enough for the temperature to come into equilibrium with the heat from the heat lamp, all three materials will have nearly the same temperature. Slight differences might be explained by the presence of convection currents in the fluids but not in the sand, which would increase cooling through the walls of the containers.

    d) The water should cool off most slowly. This is because water has an extremely high specific heat capacity, higher than almost any other substance. The specific heat capacity of a substance measures the heat energy needed to raise a unit mass of the material by one degree Celsius. Offsetting this effect are two different, but less important effects: (1) The water will undergo convection, because cooling is concentrated near the walls of the container, and that tends to bring warm interior water to the walls, thereby increasing heat conduction out to the air surrounding the container. (2) The bulk density (mass per unit volume) of the sand is about twice that of the water, so even though the specific heat capacity of the water is greater than that of the sand, there is more mass to cool down. The effect of greater specific heat capacity will overshadow these other two effects.

    e) Because of the high heat capacity of water, areas located near large bodies of water (the oceans, and large lakes) tend to have a smaller range of temperatures than areas far away from large bodies of water.

Teacher Commentary

NOTES

INVESTIGATING CLIMATE AND WEATHER

Part C: Creating Climate Clues

1. To help you learn about climates, each group in your class will specialize in researching a particular climate type.

   Choose a particular climatic zone on the map.

   You are going to prepare a set of six clues that your classmates will need to use to guess your climate type.

   Clues that you can give about your climate can include:
   - graphs of monthly temperature and precipitation for the climate;
   - descriptions of what people are wearing on a typical day;
   - nearby bodies of water;
   - well-known landforms;
   - typical animals that live there;
   - crops that are grown in the area.

2. Use all the resources you have available to conduct your research and make up your clues.

3. When all the clues are ready, each group in the class gets the chance to present its clues to the other groups.

4. To participate, someone from another group must raise his or her hand. That person then gets a chance to guess what the climate is. The person trying to guess must also give a good reason why he or she thinks your clue fits with a certain climate.

5. Continue until all groups have a chance to present their climate clues.

   a) Which clues were the most useful in guessing the climates, and why?

   b) Which resources were most useful, and why?

Teacher Commentary

Part C: Creating Climate Clues

1. This part of **Investigation 6** gives students a chance to conduct research on one particular climate type. Groups also prepare climate clues that can allow them to apply immediately what they have learned.

2. Provide your students with resources to help them to research the climate they have selected. The *Investigating Earth Systems* web site contains links that can help students get started with their research.

3. Encourage your students to use as much creativity as possible in their presentation of climate clues. They may want to use verbal clues, posters, costumes, PowerPoint presentations, or other methods of presentation. Emphasize to your students, as they play the "climate clues game," that the respondents must be able to link the clue with an aspect of the climate. For example, an umbrella and shorts might be a clue to a tropical rain forest, because it is hot and wet there.

4. You may want to keep score, perhaps on a flip chart, as the game is played. This will allow all students to participate in the game.

5. Answers to these questions will vary.

Assessment Tool
Investigation Journal–Entry Evaluation Sheet
Use this sheet to help students learn the basic expectations for journal entries that feature the write-up of investigations. It provides a variety of criteria that both you and your students can use to ensure that their work meets the highest possible standards and expectations. Adapt this sheet so that it is appropriate for your classroom, or modify the sheet to suit a particular investigation.

Assessment Tool
Student Presentation Evaluation Form
Use the **Student Presentation Evaluation Form** as a simple guideline for assessing presentations. Adapt and modify the evaluation form to suit your needs. Provide the form to your students and discuss the assessment criteria before they begin their work.

Investigation 6: Climates

Digging Deeper

WEATHER AND CLIMATE
The Difference between Weather and Climate

Weather is the state of the atmosphere at a particular place and time described in terms of temperature, air pressure, clouds, wind, and precipitation. Climate is the long-term average of weather. It is observed over periods of many years, decades, and centuries. In many areas of the United States, the daily high temperature or the daily low temperature can vary by as much as 30°F from one day to the next. In contrast, the average temperature for a whole year seldom varies by more than 1°F.

Factors That Determine the Climate

The two most important factors in describing the climate of an area are temperature and precipitation. The yearly average temperature of the area is obviously important, but the yearly range in temperature is also important. Some areas have a much larger range between highest and lowest temperature than other areas. Likewise, average precipitation is important, but the yearly variation in rainfall is also important. Some areas have about the same rainfall throughout the year. Other areas have very little rainfall for part of the year (the dry season) and a lot of rainfall for the other part of the year (the wet season).

As You Read...
Think about:
1. What is the difference between weather and climate?
2. What factors determine the climate?
3. How does average yearly temperature vary with latitude?
4. How does precipitation affect vegetation?
5. Why can local climate vary over very short distances?

Teacher Commentary

Digging Deeper

This section provides text and photographs that give students greater insight into the difference between weather and climate. You may wish to assign the **As You Read** questions as homework to help students focus on the major ideas in the text.

As You Read...

1. Weather refers to the atmospheric conditions at a particular time, day to day. Climate refers to the long-term average of weather, and the characteristic patterns of weather, in a particular region of the Earth, over years, decades, centuries, or longer.

2. The most important factors that determine climate include latitude, elevation, proximity to large bodies of water, and continental position (on the west side of a continent versus on the east side of a continent).

3. Generally, areas at low latitudes have higher average temperatures, and areas at high latitudes have lower average temperatures. The range of temperature variation throughout the year, however, depends on both latitude where the area is located in relation to the ocean.

4. The amount of precipitation an area receives affects the types of plants found in that area. Areas that receive moderate to high amounts of precipitation throughout the year generally are heavily forested, whereas areas with less rainfall are mainly grasslands or deserts. In parts of the world where precipitation varies greatly with season, areas that have most rainfall during the warm growing season will be more heavily vegetated than areas that have most rainfall during the cold season.

5. Local climate can vary over short distances because of position relative to local topography (mountain ranges; steep land slopes) and to local large bodies of water.

Assessment Opportunity

You may wish to rephrase selected questions from the **As You Read** section into multiple choice or "true/false" format to use as a quiz. Use this quiz to assess student understanding and as a "motivational tool" to ensure that students complete the reading assignment and comprehend the main ideas.

About the Photos

The lakeside city in the left photo on page C65 of the student text has a milder climate (cooler summers and milder winters, on average) than areas farther inland. The site in the right photo is located in a region of high elevation and rigorous climate, with winter snowfall greater than summer melting, leading to the existence of high ice fields and local glaciers draining the ice fields.

INVESTIGATING CLIMATE AND WEATHER

The average temperature in an area depends mainly on the latitude. Generally, areas near the Equator have high average temperatures, and areas nearer the North and South Poles have lower average temperatures. The range of temperature, however, depends more on where the area is located in relation to the ocean. Areas where winds usually blow from the ocean have a smaller range of temperature than areas far away from the ocean, in the interior of a continent. That is because water has a much greater heat capacity than rock and soil, as you saw in your investigation. It takes much more heat from the Sun to warm up water than it takes to warm up rock and soil. Likewise, water cools off much more slowly than rock and soil on cold, clear nights and in the winter.

Climate and Vegetation

The plant community in an area is the most sensitive indicator of climate. Areas with moderate to high temperatures and abundant rainfall throughout the year are heavily forested (unless humans have cleared the land for agriculture). Areas with somewhat less rainfall are mainly grasslands, which are called prairies in North America. Humans have converted grasslands into rich agricultural areas around the world. Even in areas with high yearly rainfall, trees are scarce if there is not much rainfall during the warm growing season. As you know, regions with not much rainfall and scarce vegetation are called deserts, or arid regions. Areas with somewhat greater rainfall are called semiarid regions. The major problem with using semiarid or arid regions for agriculture is that ground water is removed to irrigate crops. In many cases, the removal of

Teacher Commentary

About the Photos

In the temperature forest shown in the top photo on page C66 of the student text, temperatures in the warm season are conducive to growth of trees, and rainfall is abundant in both the cold season and the warm season. In the grasslands shown in the bottom photo, rainfall is unevenly distributed throughout the year; only low herbaceous plants, not trees, can survive the dry season.

Investigation 6: Climates

ground water exceeds the rate at which it is naturally replaced (by precipitation that reenters the ground water system). As a result, water resources are lost, water levels in wells fall, and less ground water flows to recharge streams and rivers, which has other impacts on the Earth system. Another problem with semiarid regions is that when humans use them for agriculture, the loss of natural vegetation can cause the areas to become deserts.

Microclimate

It is easy to understand how climate can vary over very large areas, because of slight changes in temperature or rainfall. Climates can also vary over very short distances. Local differences in climate are described by the term "microclimate." Differences in microclimate might explain some of the differences in weather from place to place you likely observed in the first investigation in this module.

Sometimes, low-lying areas are colder at night than higher ground nearby. On clear nights, the ground is chilled as its heat is radiated out to space. The cold ground then chills the air near the ground. The chilled air is slightly denser than the overlying air, so it tends to flow slowly downhill, in the same way that water flows downhill. The cold air "ponds" in low areas. These are places where the first frosts of autumn are earliest and where the last frosts of spring are latest. If you ever have a chance to plant fruit trees, plant them on the highest ground around!

In hilly areas, north-facing slopes get less sunshine than south-facing slopes. Local temperatures on the north-facing slopes are colder than on south-facing slopes in both summer and winter. In areas with winter snows, the snow melts much later on north-facing slopes.

Teacher Commentary

About the Photos

The photo on page C67 of the student text shows a mountainous area with an almost glacial climate. Winter snowfall persists in areas sloping to the north (to the left in the photo), where the snow does not melt away completely until late in the summer melting season. Only a slight cooling of the climate would be needed for small glaciers to form in the depressions on the north-facing slopes. There appear to be glaciers in the higher elevations in the distance.

INVESTIGATING CLIMATE AND WEATHER

Review and Reflect

Review

1. Which climate type has the most rainfall? Which has the least?
2. Which climate type has the shortest growing season? Which has the longest?
3. How and why do oceans and continents affect climate?

Reflect

4. Does climate influence human population size in an area? If so, how?
5. Imagine that the climate in your area changed suddenly.
   a) What would have to be done to homes and other buildings?
   b) How would this affect what you wear?
   c) How would this affect what you eat?
   d) How would it affect what you could grow in a garden?
   e) How would it affect transportation?
   f) How would it affect the work people do?
   g) How would it affect what you do for recreation?

Thinking about the Earth System

6. On your *Earth System Connection* sheet, note how the things you learned in this investigation connect to the geosphere, hydrosphere, atmosphere and biosphere.
7. Global warming would cause sea level to rise. How would this affect other Earth systems? What would be the effect? Higher sea level means a higher base level for rivers. How might this affect erosion by rivers?

Thinking about Scientific Inquiry

8. How did you use questions to answer by inquiry in this investigation?
9. How is a hypothesis different from a guess?
10. What factors must be considered when designing a "fair" experiment?

Teacher Commentary

Review and Reflect

Review
Your students should have gathered enough evidence to provide a reasonable answer to the **Key Question**. Emphasize that it is evidence that counts. They may have difficulty in understanding what constitutes evidence as opposed to opinion or inference. Use this opportunity to clarify both the nature and importance of evidence in scientific inquiry.

1. The tropical rain forest, monsoon climate has the most rainfall. Desert climates have the least rainfall.

2. Cold, high-latitude climates have the shortest growing season (zero!). Tropical rain forests have the longest growing season (all year long!).

3. Because of the high heat capacity of water, areas located near large bodies of water, like the ocean or large lakes, tend to have a smaller range of temperatures than areas far away from large bodies of water.

Reflect
4. Yes, in general climate affects the size of the human population in an area. Humans need a ready supply of food and water. Humans are able to modify their environments (e.g., by irrigating land, transporting water long distances with aqueducts, or building structures to protect them from the cold) to make otherwise inhospitable climates livable. In general, however, large population centers tend to be located in regions without great extremes in climate. This is less true nowadays, of course, than in the past.

5. Answers to these questions will vary depending on where you live and on how the climate is assumed to change. To make the challenge more manageable for you, you may want to specify to students how the climate is changing; i.e., suppose it got warmer or cooler, or wetter or drier.

Thinking about the Earth System
The knowledge students will have gained from **Investigation 6** presents a good opportunity for looking at the bigger picture of the Earth System. Help your students to make as many connections as they can between their understanding of the factors that determine climate and the components of the Earth System. Remind them to look at the Earth System diagram on page Cviii of the student text.

6. Climate affects weathering rates, and therefore the geosphere. The location of land masses, determined by plate tectonics (geosphere), also affects climate. The amount of precipitation, or proximity to a large body of water (hydrosphere) affects climate. And the distribution of different kinds of plants and animals (biosphere) is determined largely by climate.

7. Students may have difficulty with this question. Particularly if they have not been introduced to the concept of base level. A rise in sea level would cause flooding of low-lying coastal areas. Coastal ecosystems (the biosphere) would need to shift in position. Or, if sea-level rise is too fast, coastal ecosystems would be disrupted and altered. River valleys would be flooded by the rise in base level. The estuaries thus formed would constitute new ecosystems. Less sediment would reach coasts, accelerating coastal erosion.

Teaching Tip
Encourage students to reflect on the investigations they have completed and connect what they have discovered to the Earth system. Remind students to enter any new connections that they have found on the *Earth System Connection* sheet in their journals. This might be a good time for students to review the entries they have made on their sheets. They are approaching the end of the module. Are all the connections they have made entered on the sheet? Which connections are they having difficulty making?

Thinking about Scientific Inquiry
Help students to think about the inquiry processes they used in **Investigation 6**.

8. Students came up with ideas about the weather where they live to help them decide which climatic region they live in. Students also determined what factors are used to determine the climate of an area.

9. A hypothesis is a testable statement or idea about how something works. It is based on what you think that you know or understand already. A hypothesis is never a guess. You test a hypothesis by comparing it to observations or data that already exist or that can be gathered in the future. A hypothesis forms the basis for making a prediction, and is used to design an experiment or observation to find out more about a scientific idea or question. Guesses can be useful in science, but they are not hypotheses.

10. For a test to be fair, certain variables must be controlled (i.e., remain the same throughout the experiment).

Assessment Tool
Review and Reflect Journal–Entry Evaluation Sheet
Use the general criteria on this evaluation sheet for assessing content and thoroughness of student work. Adapt and modify the sheet to meet your needs. Consider involving students in selecting and modifying the criteria for evaluating their reflections on **Investigation 6**.

Teacher Commentary

NOTES

Teacher Review

Use this section to reflect on and review the investigation. Keep in mind that your notes here are likely to be especially helpful when you teach this investigation again. Questions listed here are examples only.

Student Achievement

What evidence do you have that all students have met the science content objectives?

Are there any students who need more help in reaching these objectives? If so, how can you provide this?

What evidence do you have that all students have demonstrated their understanding of the inquiry processes?

Which of these inquiry objectives do your students need to improve upon in future investigations?

What evidence do the journal entries contain about what your students learned from this investigation?

Planning

How well did this investigation fit into your class time?

What changes can you make to improve your planning next time?

Guiding and Facilitating Learning

How well did you focus and support inquiry while interacting with students?

What changes can you make to improve classroom management for the next investigation or the next time you teach this investigation?

Teacher Commentary

How successful were you in encouraging all students to participate fully in science learning? _____

_____

How did you encourage and model the skills values, and attitudes of scientific inquiry? _____

_____

How did you nurture collaboration among students? _____

_____

Materials and Resources

What challenges did you encounter obtaining or using materials and/or resources needed for the activity? _____

_____

What changes can you make to better obtain and better manage materials and resources next time? _____

_____

Student Evaluation

Describe how you evaluated student progress. What worked well? What needs to be improved? _____

_____

How will you adapt your evaluation methods for next time? _____

_____

Describe how you guided students in self-assessment. _____

_____

Self Evaluation

How would you rate your teaching of this investigation? _____

_____

What advice would you give to a colleague who is planning to teach this investigation? _____

_____

NOTES

Teacher Commentary

INVESTIGATION 7: EXPLORING CLIMATE CHANGE

Background Information

Glaciation

A glacier is a large body of ice and snow, resting on land (or, if floating in the ocean, then anchored to land at a number of points) and flowing by internal deformation under its own weight. That carefully worded definition emphasizes that a glacier is different from a large block of ice sliding down a sloping tabletop. Glacial ice must accumulate on the land surface to a minimum thickness of a few tens of meters before the pressure is great enough to allow the ice to behave plastically: i.e., to flow by internal deformation under its own weight. This effect of plasticity is somewhat like that of "silly putty," which acts as a solid when it is struck with a hammer, but when left on the tabletop as a lump, sags downward by flowing under its own weight. In terms of how glacial ice flows, you can think of it as an extremely viscous liquid.

In an approximate way, the flow of ice in a valley glacier is similar to the flow in a river. The vertical distribution of ice velocity is qualitatively similar to the velocity distribution in a river. The maximum velocity is at the surface of the glacier, and the minimum velocity is at the base. The reason is that the downglacier driving force—the weight of the glacier—acts throughout the ice mass, whereas the counterbalancing force—the friction at the base—acts only at the base. In a continental-scale ice sheet, the glacier flows even though it does not rest on a sloping land surface. The reason simply is that the elevation of the glacier surface is greater in the center of the ice sheet than around its edges.

There are two different components of glacier velocity: flow by internal deformation, and solid-body motion by slip at the glacier base. One important way of classifying glaciers is by the temperature of the ice at the base: there are cold-based glaciers, in which the ice at the base is below the melting temperature, and there are warm-based glaciers, in which the ice at the base is at the melting temperature. Only glaciers in which the ice at the base is at its melting point can undergo basal slip. In such glaciers, a thin film of lubricating water, typically no more than a few millimeters thick, develops in two ways: heat flow from the interior of the Earth, and frictional heat generated by the friction of sliding. If the ice at and near the base of the glacier is at the melting temperature, then all of the heat added to the base of the glacier goes toward melting of ice. Consequently, it can't be conducted upward, because there is no vertical gradient of temperature, as is the case for cold-based glaciers. (Keep in mind that heat conduction can happen only when the temperature of the material is different from place to place in the material.) Cold-based glaciers are frozen fast to their rock beds, and do little or no geological work. Warm-based glaciers, on the other hand, are very effective in eroding, transporting, and depositing rock and mineral particles, large and small.

There are two related but different aspects of glacial movement. New glacial ice is formed in the zone of accumulation, moves downglacier, and is melted (or calved into the ocean) in the zone of ablation. On the other hand, the terminus of the glacier may advance, retreat, or stay in about the same position, as glacial ice moves to the terminus and melts (or is calved).

The activity of a glacier is described by its regimen. A glacier with an active regimen has large values of both accumulation and ablation; such glaciers are usually fast moving. A glacier with an inactive regimen has small values of both accumulation and ablation; such glaciers are usually slow moving. High-latitude glaciers, like the northern part of the Greenland ice sheet or the central part of the Antarctic ice sheet, have a very inactive regimen, because both winter snowfall and summer melting are slight. The mid-latitude Pleistocene ice sheets, in contrast, must have had a very active regimen.

On the other hand, the balance between accumulation and ablation is described by its economy. In a glacier with a positive economy, accumulation is greater than ablation, and the glacier grows in volume (and the terminus usually advances). In a glacier with a negative economy, ablation is greater than accumulation, and the glacier shrinks in volume (and the terminus usually recedes). There are thus four combinations of regimen and economy that can characterize a glacier, because regimen and economy are independent of one another.

Ice Ages

We think of the existence of major ice sheets in high-latitude land areas as normal, because in the recent past there have always been high-latitude ice sheets. If we look at all of Earth's history, however, we find that for long periods—for much the greater part of Earth history, in fact—there were no ice sheets. There have been four distinct periods of glaciation, called glacial periods or simply glacials, in North America in the last 1.6 million years, with briefer periods, called interglacial periods or simply interglacials, during intervening times. The most recent glacial period ended about 10,000 years ago.

The time between 1.6 million years ago and 10,000 years ago is called the Pleistocene Epoch (one of the standard time divisions of Earth history). In addition to the Pleistocene Epoch, there seem to have been at least three other periods of glacial activity, at about 2 billion, 600 million, and 250 million years ago. As the most recent period of glaciation, the Pleistocene Epoch is interesting because the features of many present landscapes are a reflection of the work of Pleistocene glaciers. Ice covered approximately 22,000,000 km^2 of the Earth's surface during the height of the Pleistocene glaciations. In comparison, the areas of the present ice sheets in Greenland and Antarctica are 2,175,600 and 14,200,000 km^2, respectively.

The Causes of Ice Ages

Many theories for the causes of the ice ages have been proposed. Clearly, glaciers form because the Earth's climate changes in such a way as to make possible the development of continent-scale ice sheets. But that just pushes the question back further: what causes the climate to change in that way? Keep in mind that it's not just a matter of the climate becoming colder. To be conducive to glaciation, the climate must be such that the conditions for winter snowfall are enhanced and/or the conditions for summer melting are lessened. Temperature is a major factor, but precipitation is also important. A warmer climate means more summer melting, but it also is conducive to greater winter snowfall.

Two separate aspects of glaciation need to be considered when trying to develop a theory for glaciation. On the one hand,

Teacher Commentary

during a time in Earth history when ice sheets are present, the evidence is incontrovertible that the ice sheets expanded and contracted regularly on time scales of the order of several tens of thousands of years. On the other hand, there have been only a few such periods of fluctuating ice sheets during Earth history; at other times, there were no major volumes of glacier ice on Earth.

Because glaciers can form only on land, we know that for ice sheets to exist, there must be large land masses at high latitudes. By the operation of plate tectonics, the continents have shifted their positions greatly through geologic time. The geologic record shows clearly that at times of major glaciation, large continents were located at high latitudes. The Antarctic ice sheet did not begin to develop until the continent of Antarctica moved into high latitudes, long after the original breakup of the supercontinent of Pangea, of which Antarctica was a part. Plate movements can account for the major periods of glaciation in Earth history, but they do not explain the repetitive growth and shrinkage of ice sheets during those periods.

It is now almost universally accepted among climatologists that long-term climate change leading to glaciation and deglaciation is linked to changes in insolation (solar radiation received by the Earth) related to variations in the Earth's orbital parameters. These cycles of solar radiation are called Milankovitch cycles, after the Serbian scientist who first systematized the effect of the Earth's orbital parameters on insolation cycles. The Milankovitch cycles are discussed in greater detail later in this Teachers Edition in the **Background Information** section of **Investigation 8**.

More Information...on the Web

Go to the *Investigating Earth Systems* web site www.agiweb.org/ies for links to a variety of web sites that will help you deepen your understanding of content and prepare you to teach this investigation.

Investigation Overview

In **Investigation 7**, students answer questions about different "climate clues" to understand how scientists are able to study how climate has changed through time. Students consider what the presence of warm-water shell fossils in present-day Greenland suggests about the past climate of Greenland. Other evidence examined includes a fossilized impression of a banana found in Oregon, dinosaur fossils found near Antarctica, ice cores, cave paintings from the Sahara Desert, and tree growth rings. **Digging Deeper** explains the use of proxies for measuring climate change and looks at the causes of climate change.

Goals and Objectives

As a result of **Investigation 7**, students will understand how scientists use evidence to determine possible climatic changes in the recent and distant past.

Science Content Objectives

Students will collect evidence that:
1. Climate changes over time.
2. Climate trends may be natural, or affected by human activity.
3. Fossils, ice cores, tree rings, and ocean bottom cores can provide evidence of past climates.
4. A great deal of evidence about climate must be examined to arrive at any conclusions about the direction of climate change.

Inquiry Process Skills

Students will:
1. Analyze climate information, over both the short term and the long term.
2. Draw conclusions about climatic conditions from a variety of sources.
3. Share findings with others.

Connections to Standards and Benchmarks

In **Investigation 7**, students explore how climate has changed in the recent and distant past. These observations start them on the road to understanding the National Science Education Standards and AAAS Benchmarks shown below.

NSES Links

- Fossils provide important evidence of how life and environmental conditions have changed.

- Global patterns of atmospheric movement influence local weather. Oceans have a major effect on climate, because water in the oceans holds a large amount of heat.

- Fossils provide important evidence of how life and environmental conditions have changed.

Teacher Commentary

AAAS Links
- Climates have sometimes changed abruptly in the past as a result of changes in the Earth's crust, such as volcanic eruptions or impacts of huge rocks from space.

- Even relatively small changes in atmospheric or ocean content can have widespread effects on climate if the change lasts long enough.

- Heat energy carried by ocean currents has a strong influence on climate around the world.

Preparation and Materials Needed

Preparation
You will want to collect information about the current climates of Greenland, Oregon, Antarctica, and the Sahara Desert. Students need to understand the present-day climates of these areas in order to understand that the clues indicate climate change over time. Additional information students will most likely need includes where bananas grow, and information about the climate needed for dinosaurs to survive. The *Investigating Earth Systems* web site contains resources that will be helpful. You can also gather books, CD-ROMs, etc.

Materials
- resources on global climate change (books, CD-ROMs, Internet access, etc.)

Investigating Climate and Weather

Investigation 7: Exploring Climate Change

Investigation 7:
Exploring Climate Change

Key Question
Before you begin, first think about this key question.

What evidence suggests that climate has changed in the past?

Think about what you have learned so far. Is global climate changing? How do you think scientists learn about what the global climate was like before weather data were recorded? What "climate clues" are out there?

Share your thinking with others in your class. Keep a record of the discussion in your journal.

Materials Needed

For this investigation your group will need:

- resources on global-climate change (books, CD-ROMs, Internet access, etc.)

Investigate
1. Studying weather data is only one way of learning about climate and how it changes. Paleoclimatologists are climatologists who study evidence from the past (ice cores, ocean bottom cores, tree rings, rocks, and fossils, among others) to find out more about climate in the past.

Teacher Commentary

Key Question

Give students five minutes to respond in their journals to the **Key Question**, "What evidence suggests that climate has changed in the past?" You may want to write the question on the blackboard.

Discuss students' ideas about how scientists learn how climate has changed over geologic time. Ask them to explain their thinking. If you want students to complete **Investigation 7** within one class period, keep this warm-up brief.

Student Conceptions about Global Climate Change

Students are likely to have heard about global climate change through the popular media. Having completed the previous investigations, students now know the difference between climate and weather. However, they may not have a clear sense of the tools that scientists use to determine how climate has changed through time. You may also want to explain to students that the word "past" has different connotations in geology than they may be familiar with: when geologists study "past" climates, they can be looking at climates from hundreds of millions, or even billions, of years ago.

Answer for the Teacher Only

Many lines of evidence demonstrate climate change in the past. See **Background Information** for **Investigation 7** for more detail.

> ### About the Photo
> The valley in which this dam is now located was once filled with glaciers during the most recent glacial period.

Assessment Tool

Key–Question Evaluation Sheet
Use this evaluation sheet to help students understand and internalize basic expectations for the warm-up activity.

Investigate
Teaching Suggestions and Sample Answers

> **Teaching Tip**
>
> Students may not realize that the Earth has left a record of the past, including climate and weather patterns, that can be interpreted by geologists. They may not know that evidence of these patterns can show up in many different ways, including animal and plant fossils, and deposits left by glacial ice and deposits on the bottom of the ocean.
>
> In **Investigation 7**, students use a number of inquiry processes. It is important to draw their attention to these processes. Encourage them to consider carefully the processes they use, and when and how they use them. In the following—and final—investigation, students will be deciding for themselves how and when to use inquiry processes. Use this investigation to prime them for this final task.

Teacher Commentary

NOTES

INVESTIGATING CLIMATE AND WEATHER

Look at the climate evidence that follows.

Also look at the climate map of the world in the previous investigation.

Read the information about the items in the pictures (fossils, tree rings, etc.) to find out more about climate in the past.

Discuss the questions with your group.

Keep a record of your group's discussion in your journal.

Climate Evidence 1

Teacher Commentary

NOTES

Investigating Climate and Weather

Investigation 7: Exploring Climate Change

Fossil shells, found on the coast of Greenland, show that certain species of warm-water mollusks lived there about 8500 years ago.

a) What is the climate like in Greenland today?

b) What clues do you think these fossils give about climate changes in Greenland?

c) What ideas do you have about what might have caused climate changes in Greenland?

Inquiry

Considering Evidence

Scientific judgments depend on solid, verifiable evidence. Claims by scientists that the global climate is changing must be supported by evidence. You must be certain about the reliability of any evidence that you use. You must also know how the accuracy of the evidence can be checked.

Climate Evidence 2

A fossilized impression of a banana, 43 million years old, was found in the state of Oregon. Find some of the places on the world climate map where bananas grow in the world today.

a) What is the climate like in Oregon today?

b) What clues can the fossil give you about climate changes in Oregon?

c) What ideas do you have about what might have caused climate changes in Oregon?

Teacher Commentary

Climate Evidence 1

a) The climate of Greenland today is arctic to subarctic, with cool summers and cold winters. Most of Greenland, except along the southern coasts, is covered by glacial ice. Generally, it is very dry in northern Greenland. Southern Greenland has much higher annual precipitation.

b) The fact that warm-water mollusks lived in Greenland 8500 years ago suggests that the climate of Greenland was once much warmer. Specifically, coastal waters must have been warmer.

c) Answers will vary. The climate changes in question happened within the Earth's postglacial history. The last glacial period, when there were large ice sheets in North America and Europe, came to an end between 20,000 and 10,000 years ago. Climate is known to have changed substantially since then on time scales of centuries and millennia. The reasons for such shorter-term changes in climate are not yet well understood. Postglacial climate change in coastal Greenland must have been related to changes in ocean circulation in the North Atlantic.

Climate Evidence 2

Teaching Tip

Bananas grow in tropical areas where the climate is sunny and hot, and there is plenty of rain. Bananas are native to Asia, but today they are cultivated in tropical areas in Central America, South America, and Africa as well.

a) Oregon (at least that part of Oregon west of the Cascades Mountains) today is found in the Marine West Coast climatic region, meaning it has no dry season and precipitation exceeds 1.2 in. each month of the year. The average temperature of the coolest month of the year does not fall below 32°F, and does not rise above 64.4°F. The average temperature of the warmest month of the year is greater than 50°F.

b) The fossilized banana found in Oregon indicates that the climate of Oregon was once much warmer than it is today.

c) Probably the most important factor in climate change in Oregon over the past 50,000,000 years is the drift of the various continents after the breakup of the supercontinent, Pangea. At the time that bananas grew in Oregon, the area was located at a much lower latitude than now.

Climate Evidence 3

This particular photo shows an excavation of dinosaur bones at Dinosaur National Monument. The person in the picture indicates the relative size and variation in the bones.

a) Antarctica today is found in the "ice cap" climatic region. This means that the climate is cold and icy. The average temperature of the warmest month of the year is below 50°F.

b) Dinosaurs lived in climates that were warm and humid, with abundant vegetation.

c) The presence of dinosaur fossils on Antarctica indicates that the climate was much warmer.

Climate Evidence 4

This photo comes from an ice core from the Greenland Ice Sheet Project (GISP-2) taken from a depth of 1850 meters. The dark layers are actually clear in nature. They appear dark because the photographer used a black background to shoot the photograph. They represent winter layers. The light grey layers are summer layers. The difference between the clear and the darker, light grey layers comes from the concentration of dust trapped in the ice. There is always dust falling out of the air. Snow nucleates around dust particles in the atmosphere, and dust just falls out of the atmosphere and is trapped in the ice. In the summer, the sun comes out and sublimates off some of the snow, concentrating the dust, and leaving a dust layer (light grey band) in the ice. Conversely, this process does not occur during the winter and so the dust is less concentrated within the ice. As a result, the winter layer appears clearer (darker in the case of this photograph).

a) Two complete lighter layers, and one complete darker layer, and parts of two other darker layers.

b) Students may have trouble answering this question unless they are first introduced to the concept of seasonal layering in ice cores (presented in the **Digging Deeper** section of this investigation). These visible layers are the result of seasonal changes that took place as snow accumulated to form this glacier ice. The layers provide a time marker that can be used when analyzing the ice and interpreting the resulting data and age of glaciers.

Climate Evidence 5

a) Rainfall in the Sahara is sparse, with an average annual total of less than 5 in. (12.7 cm). The temperature range can be extreme. Daytime temperatures are very high; the world's highest recorded temperature in the shade is 136°F/58°C in September of 1922 in Azizia, Libya. Temperature drops radically at night, and a common diurnal range is 86°F (30°C). Between December and February, nighttime temperatures can fall below freezing. Lastly, the Sahara is located within the trade winds belt and the region regularly experiences strong winds that blow constantly from the northeast.

b) In order for large animals to have survived in the Sahara, the region must have been able to offer a source of food and water and shelter. More rain would have supported grasslands, forests, as well as rivers. Cooler day temperatures would have made the environment more hospitable for animals like antelopes, giraffes, elephants, crocodiles, hippopotami and rhinoceroses, and especially people.

Teacher Commentary

Climate Evidence 6

The tree growth rings shown are that of a Douglas fir tree from northeastern Arizona. The oldest living Douglas fir tree is 1275 years old. This tree was approximately 580 years old when it was harvested. It is not necessary to kill a tree to date it. Instead, scientists can take a small core of the tree and use that to date it. Using this technique, revered trees like the bristlecone pine, which is not typically harvested, can be scientifically studied without doing any lasting harm to the tree.

a) Students should not be asked to count all of the tree rings. Instead, the arrow marking "550" can be used as a reference, and they can then count the number of rings inside of that mark, add it to 550 and then arrive at the age of the tree. There are approximately 30 rings inside of the 550 mark, making a total of 580 rings or a tree age of 580 years.

b) Once again, students may have difficulty answering this question without necessary background information presented below. The tree rings are thinner in the outer part of the tree than in the inner part, indicating that climate was getting progressively colder and/or drier.

Teaching Tip

Great care must be taken when the succession of thicknesses of a tree's annual growth rings is used for paleoclimate study. Ring width varies as a function of a number of factors. During the life of the tree, the early robust growth results in relatively thick rings in the inner part of the trunk, whereas slower growth later in the life of the tree results in relatively thin rings. This effect must be taken into account by comparing the succession among many different trees in a given area. Local effects having to do with shading of one tree by another, or by death of a neighboring tree, thus abruptly increasing the sunlight available to a given tree, also necessitates the use of more than a single tree in paleoclimate studies. The predominant effects on ring width in trees of a given local area, however, are temperature and precipitation.

Cooler temperatures during the growing season cause a slowdown in the metabolic processes in the tree, and thus less addition of new wood. A related effect is that in a cool year the growing season is shorter. Rate of growth of new wood is also strongly affected by availability of soil moisture available to roots. One of the principal problems in interpreting paleoclimate on the basis of tree rings lies in separation of the temperature effect and the precipitation effect.

Assessment Tool

Investigation Journal–Entry Evaluation Sheet
Use this sheet to help students learn the basic expectations for journal entries that feature the write-up of investigations. It provides a variety of criteria that both you and your students can use to ensure that their work meets the highest possible standards and expectations. Adapt this sheet so that it is appropriate for your classroom, or modify the sheet to suit a particular investigation.

INVESTIGATING CLIMATE AND WEATHER

Climate Evidence 3

Fossils found near Antarctica include a meat-eating dinosaur that lived 200 million years ago. The dinosaur preyed on small animals that, in turn, fed on lush plants.

a) What is the climate like in Antarctica today?

b) What did the climate need to be like for the dinosaur to survive?

c) What clues can the fossils give about climatic changes in the area around Antarctica?

Climate Evidence 4

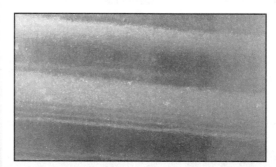

Ice cores can be taken wherever very deep ice exists. Some of these ice cores have sampled ice that is more than 400,000 years old. Ice traps air and dust in tiny bubbles in the ice. Climatologists can study these materials to find out about past climates.

a) How many darker layers and how many lighter layers can you see in this example?

b) What do the layers suggest about the climate during the time period that was sampled?

Teacher Commentary

NOTES

Investigation 7: Exploring Climate Change

Climate Evidence 5

Cave paintings like this have been found in the Sahara Desert. They have been dated as far back as 4000 B.C. Observe the kinds of animal it shows.

a) What is the climate like in the Sahara today?

b) What did the climate need to be like for the animal in the painting to live in the Sahara?

Climate Evidence 6

This picture shows the tree growth rings from a bristlecone pine tree. Some of these trees live to be as old as 4000 years. A wide ring means warm, humid weather; a narrow ring means cold, dry weather.

a) Count the number of growth rings on this example (approximately the number of years the tree has lived).

b) What evidence do you have about past climate, on the basis of the tree rings pictured?

Teacher Commentary

NOTES

Investigating Climate and Weather

INVESTIGATING CLIMATE AND WEATHER

2. When you have finished working with all of the evidence, look over what you have written about climate in the past.

   Consult any resources you have available about climate in the past to add more information to what you already know.

   a) Write down any new ideas you discover.

3. As a class, pull all your investigations together. Discuss your answers and come to a consensus as a class.

   a) List all the evidence you have been able to find that shows how climates have changed over time. Wherever possible, include the time scales involved.

   b) What evidence do you now have that shows that climate change happens slowly? What evidence do you now have that climate change can happen rapidly?

As You Read...
Think about:
1. What is a climate proxy?
2. How are ice cores used to determine past climates?
3. In your own words, explain the astronomical theory of the ice age.
4. What are two events that can affect worldwide climate over a period of a few years?

Digging Deeper

MEASURING CLIMATE CHANGE
Climate Proxies

The Earth's climate has changed greatly through the billions of years that constitute geologic time, and even in recent centuries. The study of past climates is called paleoclimatology (*paleo-* means early or past).

Something that represents something else indirectly is called a proxy. There are many proxies for past climate. They provide a lot of information, although none is perfect. Some, like kinds of past plants and animals, are easy to understand. Some important proxies, involving the chemical element oxygen, are more difficult to understand.

Ice Cores

Thin cores of ice, thousands of meters deep, have been drilled in the ice sheets of Greenland and Antarctica. They are preserved in special cold-storage rooms for study.

Teacher Commentary

2. This is an opportunity for students to look back on how they have conducted their research. Help them make any connections they can.

3. You may find that your students have an unrealistic perception of geologic time. You might find it useful to have your students make a timeline showing where the different evidence fits. This will help them answer the questions.

   a) See sample timeline below:

Years before Present	Climate Evidence	Nature of Difference in Climate
8500	Fossils of warm-water mollusks from this time are found on the coast of Greenland, which now has a cold climate and is surrounded by cold water.	Coastal waters were warmer than today. This probably resulted in a warmer climate in Greenland than is seen today, much like how warm waters of the Gulf Stream warm parts of northern Europe like the coast of Norway, which is much warmer than Greenland, but at a similar latitude.
43,000,000	A fossil of a banana found in Oregon indicates that this region once had a climate more tropical than it has today.	Past climate warmer and wetter than today.
200,000,000	Dinosaur fossils in Antarctica (now cold) indicate that once it had a climate that supported the growth of lush plants.	Past climate much warmer than today.
400,000 to present	Ice cores in Antarctica give a fairly unbroken record that goes back at least 400,000 years. They also have lighter and darker layers that indicate seasonal changes.	Antarctica has had glacier ice for at least 400,000.
4000	Cave paintings of animals that are not found in deserts today were found in the Sahara Desert. This indicates that the Sahara Desert once had wetter climate than today.	Past climate much wetter than today.
4000 to present	Tree rings from a Douglas fir tree have variable spacings, and these come in groups.	Shows changes from warm, humid climate to a colder, drier climate.

b) Evidence like the tree rings, cave paintings, and ice cores provide evidence that the climate can change over relatively rapid time scales (less than a few thousand to a few tens of thousands of years). Conversely, evidence like the fossil banana impression and the dinosaur fossils also indicate that climate can change over much longer time scales, on the order of millions to hundreds of millions of years.

> **Assessment Tools**
>
> **Journal–Entry Evaluation Sheet**
> Use this sheet as a general guideline for assessing student journals, adapting it to your classroom if desired.
>
> **Journal–Entry Checklist**
> Use this checklist as a guide for quickly checking the quality and completeness of journal entries.

Digging Deeper

This section provides text, a global temperature graph, and photographs that give students greater insight into the measurement of climate change. You may wish to assign the **As You Read** questions as homework to help students focus on the major ideas in the text.

As You Read...

1. A climate proxy is any evidence or information that can be used to represent climate indirectly. Proxies for climate include kinds of past plants and animals, data from glacier ice, tree rings, etc.

2. Scientists are able to use oxygen atoms contained in glacial ice as a proxy for air temperature above the glacier. The details are well beyond the middle-school level, however.

3. The astronomical theory of the ice ages states that small changes in the Earth's orbit trigger the advance and retreat of ice sheets.

4. Volcanic eruptions and El Niño/La Niña are two events that can affect global climate over a period of a few years.

Teacher Commentary

Assessment Opportunity

You may wish to rephrase selected questions from the **As You Read** section into multiple choice or "true/false" format to use as a quiz. Use this quiz to assess student understanding and as a motivational tool to ensure that students complete the reading assignment and comprehend the main ideas.

Investigation 7: Exploring Climate Change

Glacier ice is formed as each year's snow is compacted under the weight of the snows of later years. A slightly darker layer that contains dust blown onto the ice sheet during summer, when not much new snow falls, marks each year's new ice. The winter layer consists of cleaner and lighter-colored ice. The layers are only millimeters to centimeters thick. They can be dated by counting the yearly layers. The oxygen in the water molecules also holds a key to past climate. Scientists are able to use the oxygen atoms in the glacier ice as a proxy for air temperature above the glacier.

Past Glaciations

Ice sheets on the continents have grown and then shrunk again at least a dozen times over the past 1.7 million years. Many climate proxies make that very clear. Deposits of sediment and distinctive landforms left by these glaciers are present over large areas of North America and Eurasia. Proxies for global temperature show gradual cooling as the ice sheets form. They also then show very rapid warming as the ice sheets melt back. Intervals of relatively high temperature between glaciations are called interglacials. Past interglacials have lasted about 10,000 years. Civilization developed only within the last interglacial—and you are still in it! The graph shows the estimated global surface temperature for the last 420,000 years.

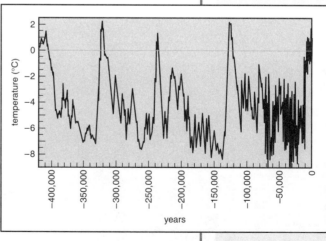

Teacher Commentary

About the Photo

The data shown in the graph on page C75 of the student text was collected from an ice core from Vostok, the Russian base in Antarctica. It dates back to 420,000 years ago.

INVESTIGATING CLIMATE AND WEATHER

This was obtained from the longest and most informative ice core. This core was taken at the Vostok station in Antarctica by a team of Russian, American, and French scientists.

Causes of Climate Change

Probably the two most important factors that determine Earth's climate are the amount of heat the Sun delivers to the planet, and also where the continents are located relative to the Equator. Continental ice sheets cannot develop unless plate tectonics cause one or more continents to be at high latitudes.

The Earth revolves around the Sun once a year. Its orbit is in the shape of an ellipse. If the Earth were the only planet, its orbit around the Sun would be almost unchanging. The pull of the other planets on the Earth causes the Earth's orbit to be much more complicated. The orbit changes slightly in several different ways. These changes occur over periods that range from about 20,000 years to about 100,000 years. They cause slight differences in how much of the Sun's heat the Earth receives in winter versus summer and at high latitudes versus low latitudes.

One theory holds that the small changes in the Earth's orbit trigger the advance and retreat of ice sheets. This

Teacher Commentary

About the Photo

The photo on page C76 shows Crater Lake. The caldera, which is now filled by the lake, was created after a large volcanic explosion at Mount Mazama. Large volcanic eruptions can send cubic kilometers of ash up into the atmosphere, affecting climate on a short time-scale.

Investigation 7: Exploring Climate Change

theory is known as the astronomical theory of the ice ages. It was first developed in the 1920s and 1930s by the Serbian astrophysicist Milutin Milankovitch (1879–1958). It was not widely accepted by the scientific community until the 1970s. Although most scientists today accept this theory, details of how the changes govern the volume of ice sheets are still only partly understood. For example, the extremely fast melting of the ice sheets, compared to the long times needed for them to form, is still a mystery.

The astronomical theory is only part of the story. Climate is known to change on time scales as short as a century or even a few years, and the cause (or causes) of these changes are still not clear.

Some violent volcanic eruptions are known to influence climate on a scale of one to two years. Substances from the volcano are blasted into the air. Some are so small that they can remain in the atmosphere for many months to a few years. While there they both absorb and reflect solar radiation. This reduces the amount of sunlight that reaches Earth's surface. A cooling of up to 1°C at Earth's surface may be observed.

El Niño and La Niña are large-scale air—sea interactions in the tropical Pacific. They can also affect the climate in many regions of the world over periods of one to two years. Changes occur in the sea-surface temperature. This affects the air pressure patterns in the tropical Pacific. As a result, storm tracks at middle and high latitudes are altered. Some regions that are usually wet have droughts. Other regions that are arid or semiarid have heavy rains.

Teacher Commentary

About the Photo

This volcanic eruption of Mount St. Helens in Washington State in 1980 is putting enormous quantities of volcanic ash high into the atmosphere, all the way into the stratosphere. The fine particles of the volcanic ash can remain in the atmosphere for months or even a few years, decreasing the solar energy received at the Earth's surface and causing temporary global cooling.

INVESTIGATING CLIMATE AND WEATHER

Review and Reflect

Review

1. Describe at least three ways that scientists can detect or measure climate change.
2. What are some of the possible causes of climate change?

Reflect

3. What kinds of evidence suggest that climates have changed over time?
4. How convincing does this evidence seem to be?
5. What further evidence is needed to answer some of the questions about climate change?

Thinking about the Earth System

6. On your *Earth System Connection* sheet, note how the things you learned in this investigation connect to the geosphere, hydrosphere, atmosphere, and biosphere.

Thinking about Scientific Inquiry

7. What are some sources of error associated with the ways of detecting climate change that are described in this investigation?

Teacher Commentary

Review and Reflect

Review

From their research, your students should have gathered enough evidence to provide a reasonable answer to the **Key Question**. Again, emphasize that it is evidence that counts. From their investigations, your students should be able to understand how climate change is measured.

1. Answers will vary but are likely to include fossil evidence, ice cores, cave paintings, or tree rings. Students should explain how each of these kinds of evidence is important by citing the examples they studied in **Investigation 7**.

2. The motion of the Earth's lithospheric plates determines where continents are located relative to the Equator and to each other. The luminosity of the Sun (the rate of emission of solar energy) has varied over geologic time, resulting in long-term climate change. The astronomical theory of the ice ages states that small changes in the Earth's orbit trigger the advance and retreat of ice sheets. Volcanic eruptions and variations in air–sea interactions (as, for example, during El Niño and La Niña conditions) can affect climate on shorter time periods.

Reflect

Give your students time to reflect on the nature of the evidence they have generated from their investigations. Again, help them see that evidence is crucial in scientific inquiry. Use this session to pull together all that students have learned in the module.

3. The kinds of evidence that support climate change over time include:
   - fossils of organisms that lived in a climate different from the current climate that exists where they are found
   - cave paintings showing animals that feed on vegetation that no longer grows where the paintings were found
   - systematic variation in the thickness of the annual growth rings of trees
   - geologic deposits and features left by glaciers in areas where glaciers no longer exist
   - data recovered from glacier ice and ocean sediment that can be used as a proxy for climate

4. Answers will vary. It is important that students understand that claims of climate change must be supported by evidence. Ask them to consider the reliability of the evidence they examined in **Investigation 7**, and whether or not its accuracy could be checked. The connection between fossil evidence and climate change will probably seem more direct and convincing to students than indirect evidence from ice cores. However, ice-core data are the principal data for interpreting climate change over the past few hundred thousand years.

5. Answers will vary. The more evidence that is available, the stronger the case that can be made to support a change in climate. However, students should understand that the evidence they have examined represents proxies for climate. They should realize that there is no way to be 100% certain.

Thinking about the Earth System
The knowledge that students will have gained from **Investigation 7** presents a good opportunity for looking at the bigger picture of the Earth System.

6. Answers will vary. A sample connection would link changes in climate brought about by volcanic eruptions (geosphere) or changes in air–sea interactions (hydrosphere and atmosphere) that result in changes in the distribution and types of organisms found in an area (biosphere).

Teaching Tip
Remind students to enter any new connections that they have found on the *Earth System Connection* sheet in their journals. Encourage them to check to make sure all the connections they have discovered are entered on the sheet.

Thinking about Scientific Inquiry
7. Here are some examples of sources of error:
   - simple miscounting of tree rings or ice-core layers
   - deformation and/or metamorphism of glacial ice, leading to uncertainty in recognizing individual annual layers in ice cores
   - dating of rocks or deposits containing fossils that give evidence of past climates

Assessment Tool
Review and Reflect Journal–Entry Evaluation Sheet
Use the general criteria on this evaluation sheet for assessing content and thoroughness of student work. Adapt and modify the sheet to meet your needs. Consider involving students in selecting and modifying the criteria for evaluating their reflections on **Investigation 7**.

Teacher Commentary

NOTES

Teacher Review

Use this section to reflect on and review the investigation. Keep in mind that your notes here are likely to be especially helpful when you teach this investigation again. Questions listed here are examples only.

Student Achievement

What evidence do you have that all students have met the science content objectives?
_____

Are there any students who need more help in reaching these objectives? If so, how can you provide this? _____
_____

What evidence do you have that all students have demonstrated their understanding of the inquiry processes? _____
_____

Which of these inquiry objectives do your students need to improve upon in future investigations? _____
_____

What evidence do the journal entries contain about what your students learned from this investigation? _____
_____

Planning

How well did this investigation fit into your class time? _____
_____

What changes can you make to improve your planning next time? _____
_____

Guiding and Facilitating Learning

How well did you focus and support inquiry while interacting with students?
_____

What changes can you make to improve classroom management for the next investigation or the next time you teach this investigation? _____
_____

Teacher Commentary

How successful were you in encouraging all students to participate fully in science learning? _____

_____

How did you encourage and model the skills values, and attitudes of scientific inquiry? _____

_____

How did you nurture collaboration among students? _____

_____

Materials and Resources

What challenges did you encounter obtaining or using materials and/or resources needed for the activity? _____

_____

What changes can you make to better obtain and better manage materials and resources next time? _____

_____

Student Evaluation

Describe how you evaluated student progress. What worked well? What needs to be improved? _____

_____

How will you adapt your evaluation methods for next time? _____

_____

Describe how you guided students in self-assessment. _____

_____

Self Evaluation

How would you rate your teaching of this investigation? _____

_____

What advice would you give to a colleague who is planning to teach this investigation? _____

_____

NOTES

Teacher Commentary

INVESTIGATION 8: CLIMATE CHANGE TODAY

Background Information

Causes of Climate Change
It is now almost universally accepted among climatologists that long-term climate change leading to glaciation and deglaciation is linked to changes in insolation (solar radiation received by the Earth) related to variations in the Earth's orbital parameters. These cycles of solar radiation are called Milankovitch cycles after the Serbian scientist who first systematized the effect of the Earth's orbital parameters on insolation cycles.

Three aspects of the Earth's movement in space account for variations in insolation over the course of a given year: eccentricity, obliquity, and precession. These phenomena are caused by the slight but significant effects of the gravitational attractions of the other planets in the solar system, as they change their positions relative to the Earth.

Eccentricity—The Earth's orbit around the Sun is an ellipse, with the Earth at one of the foci of the ellipse. The ellipse is very close to being a circle. The eccentricity of the orbit (i.e., the degree to which the ellipse is elongated, and therefore differs from a circle) varies significantly on a time scale of about 100,000 years. When the orbit is more eccentric, insolation varies more over the course of a year than when the orbit is less eccentric.

Obliquity—The Earth's axis of rotation is tilted relative to the plane of the Earth's orbit around the Sun. At present, the angle is about 23.5°. The angle changes significantly, however, within a period of about 40,000 years. At times of smaller obliquity (a small angle of inclination of the axis), insolation varies less from season to season; at times of greater obliquity, insolation varies more from season to season, particularly at high latitudes

Precession—The Earth's axis of rotation wobbles in the same way that a spinning top wobbles. With both the Earth and a spinning top, the wobble is far slower than the rotation of the body. This phenomenon is known as "precession." The period is approximately 20,000 years. Precession causes the timing of the seasons to be different relative to the Earth's orbit, and therefore to be different relative to the times when the Earth is closest to the Sun and farthest away from the Sun. At present, the Sun is closest to the Earth during the Northern Hemisphere winter (on January 5)! Don't let your students think that winter happens because the Earth is farthest from the Sun: it happens because the North Pole is tilted away from the Sun rather than toward the Sun. In another 10,000 years, when the Earth is farthest from the Sun during the Northern Hemisphere winter, our winters will be more severe—other factors being equal—just because we will be receiving less solar radiation in winter than we get now.

The three effects noted above govern changes in insolation, which are ever changing in their combination because of the three different periodicities. These changes in insolation lead to changes in season-to-season climate, in complex ways that are not yet entirely clear. What is clear is that glaciation and deglaciation are somehow linked to these changes—the time scales of glaciation and deglaciation match the periodicities of variation in insolation extremely well.

Global Warming

The issue of global warming is much in the news these days. It is generally agreed that global yearly surface temperatures, averaged around the globe, have increased slightly over the past several decades. The magnitude of this warming is not entirely certain, however, because of difficulties in intercalibration of measurements made with older instruments and those made with newer instruments and techniques, as well as the various assumptions and corrections needed to take into account the effects of urbanization (the so-called urban-heat-island effect) on temperature records from the great many weather stations located in urban areas. There is also much indirect evidence that points to global warming, especially the general retreat of glaciers in most parts of the world and the thinning of the pack ice in the Arctic Ocean.

Much of the controversy about global warming hinges upon how much of the observed warming is caused by natural effects and how much by human-induced (so-called anthropogenic) effects. It is known from the climate record of the past few millennia that global temperatures can change rapidly—by magnitudes comparable to the increase in recent decades—without human intervention. On the other hand, it is generally (although not universally) agreed that the great increase in greenhouse gases, especially carbon dioxide, is likely to cause an increase in global temperature.

Climatologists are working to develop ever more sophisticated computer models of the Earth's climate in order to simulate, and therefore predict, future global temperature and rainfall. It is not yet possible to take all of the important physical effects into account, but the computer models are in the process of continual refinement. There will never be certainty in prediction of future climate. In the view of many scientists, it comes down to a choice society has to make: we must balance the desirability of taking action now to forestall the deleterious effects of global climate change (that seem likely on the basis of what climatologists know now) against the potentially serious economic disruptions that would arise from such action.

Over the past several thousand years of Earth history, there has been a very strong positive correlation between global temperature and atmospheric carbon dioxide concentrations: when temperature has been high, carbon dioxide concentrations have been correspondingly high. Many nonscientists assume that this is an indication that the increase in anthropogenic carbon dioxide concentrations will lead to global warming. Be on your guard about such thinking, however: it's a truism in science that correlation does not prove cause and effect. The correlation between global temperature and carbon dioxide concentration could be interpreted in three ways:
- higher carbon dioxide causes higher temperatures
- higher temperatures cause higher carbon dioxide
- both higher temperatures and higher carbon dioxide are caused by some third factor

Almost all climatologists agree that climate change in recent geologic time has been caused by the astronomical effects described above. If that is so, then the warming may somehow have caused the increase in carbon dioxide! The observed correlation then becomes irrelevant to the extremely important problem of the extent to which the present anthropogenic increase in carbon dioxide will lead to global warming. Humankind is engaged in an unprecedented experiment on global climate, and unfortunately the past gives us little basis for predicting the outcome of that experiment.

Sea-Level Change

Since the beginning of the Pleistocene, the Earth has experienced changes in sea level

Teacher Commentary

of more than 100 m as a result of the growth and shrinkage of continental ice sheets. At the peak of the most recent glaciation, sea level was more than 100 m lower than it is today. Because of the gentle slopes of the continental shelves, shorelines around the world were located in positions that are now far out to sea. The Atlantic coast was more than 100 km to the east of New York City; Southeast Asia was tied to Indonesia by a strip of land; France and Britain were connected by dry land where the English Channel is now; and Alaska and Siberia were connected across the Bering Strait.

In the period between 10,000 and 6000 years ago, sea level rose over 125 m in just 4000 years. This compares to sea level along the coast of the Netherlands rising just 4 m over the last 4000 years. For the last two centuries, sea level has been rising at the rate of about 1 to 2 mm per year, or about 10 cm in the last 100 years.

A very small rise in sea level would have a large impact on the human population, because a large percentage of the population is concentrated in coastal cities. Much agriculture is also concentrated close to coasts. A 1-m rise in sea level would affect 6,000,000 people in Egypt, with 12% of the agriculture lost; 13,000,000 people in Bangladesh, with 16% of national rice production lost; and 72,000,000 in China, with tens of thousands of hectares of agricultural land lost.

In addition to direct land loss due to rising sea level, indirect factors would cause much trouble. These include changes in erosion patterns along the coasts, salinization of wells, and problems with sewage systems for coastal cities.

More Information...on the Web
Go to the *Investigating Earth Systems* web site www.agiweb.org/ies for links to a variety of web sites that will help you deepen your understanding of content and prepare you to teach this investigation.

Investigation Overview

Students use the knowledge and skills about climate and weather that they have gained throughout the module to make, and research, predictions about climate in the future.

Goals and Objectives

The purpose of this activity is to review science content and inquiry processes that have been used throughout the module. It can be used as a final assessment, as a review for a final test, or both. As a result of **Investigation 8**, students will develop a better understanding of how climate is changing and what evidence is used to support this.

Science Content Objectives

Students will collect evidence that:
1. Climate changes over time.
2. A great deal of evidence about climate must be examined to arrive at any conclusions about the direction of climate change.
3. Human activity may affect climate.

Inquiry Process Skills

Students will:
1. Make a prediction about the direction of climate change in the future.
2. Collect evidence about climate change over as long a period of time as possible.
3. Analyze the evidence they have collected on climate change.
4. Arrive at conclusions about climate predictions on the basis of evidence.
5. Communicate what they have learned about climate change to others.

Connections to Standards and Benchmarks

All the content Standards and Benchmarks that students have been working toward understanding come together in this final investigation. Remember, these are statements of what students are expected to understand by the time they complete eighth grade. What they have been doing throughout this module on climate and weather is just part of that ultimate learning outcome. Your students will have developed their understanding of some of these ideas, at least in part, but many students will require additional experiences.

As your students work through **Investigation 8**, keep these Standards and Benchmarks in mind and note the general level of understanding evident in what students discuss and do. Be especially alert to any confusion that a simple question from you might clarify, but do not attempt to "teach" these standards directly. The role here is to guide students from the ideas they already have toward a more complete understanding.

Teacher Commentary

Preparation and Materials Needed

Preparation
In **Investigation 8**, your students use what they have learned about weather and climate to investigate whether there is credible evidence that climate change is happening today. They act as researchers, gathering information from credible sources, analyzing the data they collect, and eventually turning their research into a report that they present to others.

It will be important to have as many resources available as possible. However, part of the students' role is to search for and find relevant information for themselves. In doing this, they are mirroring what scientists do.

Materials
- students' data from previous investigations
- weather data for community for past 30 years*
- resources on global climate change (books, CD-ROMs, Internet access, etc.)
- presentation materials (poster board, markers, etc.)

* The *Investigating Earth Systems* web site provides suggestions for obtaining these resources.

Investigating Climate and Weather

Investigation 8: Climate Change Today

Investigation 8:

Climate Change Today

Key Question
Before you begin, first think about this key question.

How is the global climate changing?

Think about what you have learned about climate change. Do you think that the world's climate is changing? If so, what are the prospects for the future? What will the climate be like for you, your children, and your grandchildren?

Share your ideas with your classmates. Record your thoughts in your journal.

Materials Needed

For this investigation, your group will need:

- your weather data
- weather data for your area for the past 30 years
- resources on global-climate change (books, CD-ROMs, Internet access, etc.)
- presentation materials (poster board, markers, etc.)

Investigate
1. Pull together all the weather data and climate information that you have collected over the course of the module.

Teacher Commentary

Key Question

By now, your students should have a good understanding about all the aspects of climate and weather they need to apply here, on the basis of investigations and evidence. It might be helpful to summarize and hold a brief discussion about each point, emphasizing to students that their knowledge of these aspects of climate and weather will be crucial in completing this final investigation successfully.

Student Conceptions about Global Climate Change Today

Though student understanding will be more highly developed than at the beginning of this module, most students will find it difficult to relate this question to their new knowledge. It is important that your students realize that this final investigation draws upon their accumulated understanding of climate and weather. Encourage students to approach an "open-ended" question like this with a positive attitude. It may be hard to get started, but there is more room for creativity and, ultimately, the expression of their ideas. Many of your students' initial informal ideas will be closer to commonly accepted scientific concepts and explanations. Keep in mind that understanding geologic time in relation to climate change might be difficult for some students.

Assessment Opportunity
Key–Question Evaluation Sheet
Use this evaluation sheet to help students understand and internalize basic expectations for the warm-up activity.

About the Photo
Photo is of Monument Valley, Utah, a region once covered by water. Many geological features in the southwestern United States expose a long record of geologic history that can be used to interpret the distant past.

Investigate

Teaching Suggestions and Sample Answers

Teaching Tip

Decide on the composition of student groups. Because this is the last investigation of the module, you may want to use it as an assessment tool or as a review for a final test (or both). Choose groups carefully so that there is a mix of abilities and good group dynamics. If the group dynamics are not good, change them before the end of the first part of **Investigation 8**.

Making a prediction about future climate and following the same steps that scientists follow can be exciting for the students. Make sure that they are always mindful of the need to reinterpret their knowledge about climate and weather in a way that will inform others.

Assessment Tool

Assessing the Final Investigation

Students' work throughout the module culminates with this final investigation. To complete it, students need a working knowledge of previous activities. The last investigation is a good review and a chance to demonstrate proficiency because it refers to the previous steps. For an idea on using the last investigation as a performance-based exam, see the section in the back of this Teacher's Edition. If you chose to use a scoring guide, review it with students before they begin their work.

Teacher Commentary

NOTES

INVESTIGATING CLIMATE AND WEATHER

Examine the climate of your area and compare it to the weather data you have collected.

Also compare the climate with the weather data from your area for the past 30 years.

Discuss what patterns and relationships you notice in your climate and weather data with your group.

a) What trends, if any, do you notice in the weather data over the past 30 years?

b) Do the temperatures appear to be increasing, decreasing, or staying about the same? How can you tell?

c) What is the average yearly rainfall in your area?

d) On the average, have you had more or less rainfall than one would predict for your climate over the past ten years?

e) How could you explain what you observe?

2. Now think about what the future climate might be like in the region where you live.

Think about these questions:

- Will it be much warmer or much cooler than it is now? If so, how much warmer or cooler? Or will the climate be about the same as now?

- Do you think there will be much more rainfall per year, much less, or about the same? Why do you think that?

a) Write down a prediction for what you think the climate will be like where you live 100 years from now. Also include the reasons for your prediction.

3. Discuss your predictions with other people in your class.

a) List the predictions and reasons that others have.

4. For this investigation you may decide to stay in your normal groups or to work with people who have ideas similar to yours. Whichever you decide, your next task is to look at the individual predictions and the reasons you gave for them.

In your group, develop a final prediction, with reasons, that you all agree on.

a) Write this down clearly. It will be the key starting point for your research.

Teacher Commentary

1. Answers to these questions will vary depending upon the area in which you live. Visit the *Investigating Earth Systems* web site to help you find the needed weather data for your area.

Teaching Tip
Students need to understand fully what is expected of them in this final investigation. Before they begin, walk them through **Investigation 8** in detail, emphasizing each particular component and answering any questions they may have. In doing this, help them see that they are working in the same way research scientists work. Emphasize the importance of credible evidence in research, and explain that this credibility depends upon the rigor with which the research is done. Alert students that they will need to use critical thinking skills when analyzing the information they gather.

2. This is an important part of **Investigation 8**. Students should make their predictions based on what they know about climate and weather. The predictions should have sound supporting reasons. Emphasize the role played by predictions in scientific investigations. Stress that predictions are ultimately supported, or not supported, by evidence that is valid and based on good reasoning.

    a) Student predictions will vary. Students should use the weather data they collected in **Step 1** to support their reasoning behind each prediction.

Teaching Tip
Scientists make predictions and justify them with reasons at the start of an experiment. A prediction is a statement of the expected outcome of an investigation. A prediction is never a guess. It is based on what the scientist already knows about something. The prediction, when carefully stated so that it can be proved or disproved, is called a hypothesis.

3. a) Your students can compile a master list of climate predictions. Each group should then select one prediction to research further.

4. Check the final predictions to make sure that they are feasible for students to research.

Teaching Tip
Ensure that all students pay special attention to their use of inquiry processes. They should maintain a record of their use of inquiry process throughout the investigation and be prepared to demonstrate this at the end.

Investigation 8: Climate Change Today

5. Begin by assessing what information you already have available, then discuss what further information sources you need to consult.

   Keep in mind that you need to look for data that both do and do not support your predictions, and the reasons you gave for them.

   In your search for new information, you might want to divide the tasks. For example, one person could be responsible for exploring the Internet, another for locating and investigating text books, another looking at trade books, and so on. Also, do not forget to consult each group member's journal for information.

   When you have a plan, make a schedule for the information gathering part of your research project.

   a) Record your schedule in your journal.

6. Once you have gathered all the information you can, look at what you have. In your group discuss these questions:

   - What evidence is there from all of this information that is relevant to the research prediction, and the reasons you gave for it?

   - Which parts of the evidence support the prediction, which do not support it, and which do neither but are still important?

   It may be helpful to construct a chart within which to put these various pieces of evidence. You may find an alternative way to do this. By now you have enough experience of dealing with data to decide.

   As well as using this way of sorting your information, keep in mind that you might later want to include data charts, or other forms of representation, in your final report. If one person in your group has special talents in designing these things, give this job to him or her.

7. When you have fully reviewed all your collected evidence and information, you need to analyze it against your prediction.

   This is a very important step. Here you need to decide whether your prediction is supported by the evidence, not supported, or supported somewhat but not enough to be conclusive.

Teacher Commentary

5. Although some information resources are likely to be generally available—like school library reference books, textbooks, and the students' own journals—others may have to be searched for and acquired. It is important that students be involved in the process of finding relevant information from a variety of sources. Ask them to visit their community library. There, they may be able to find scientific journals or popular magazines that review or comment upon current evidence about global warming. Some students may have access to local experts, like meteorologists at a local TV or radio station. Others may be able to tap into expertise at a local community college or university. Using a search engine on the Internet will be especially helpful for finding good information. The *Investigating Earth Systems* web site also provides a good place for students to start their research.

    a) You may want to give students a time frame around which they can build their schedules. It may be that they will need to conduct some of their research outside of class.

6. Monitor student progress. Circulate around the room to provide help if needed.

Teaching Tip

It is important for your students to have alternative methods of organizing the data they collect. One way is to use the "idea burst" shown here. Others could include a data table, a concept map, or pictures. Help students survey the kind of information they have from an organizational point of view and find the best way to record that information.

7. It is here that students need to use critical thinking skills. The information they will have gathered may vary in terms of credibility. Help students keep on track by focusing on the evidence provided by the information they are reviewing. They need to see the difference between opinion and facts in doing this.

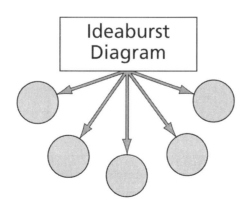

INVESTIGATING CLIMATE AND WEATHER

Together, work on this analysis and then reach agreement on how both your prediction and the reasoning behind it are reflected by the evidence. When this is complete, and all are agreed, your research phase is complete.

8. Decide on a way of presenting the results of your research for others to see and understand. This is going to be your research report. This must include:

   - your predictions and the reasons for them;
   - your evidence for or against your predictions;
   - your conclusions about whether the global climate is changing or not, based upon the evidence you have found and analyzed;
   - what inquiry processes you used in your research and how you used them;
   - how your research relates to the Earth system.

   Together, discuss how you will do this in your report. Decide what sections your report will contain, where illustrations and data representations will be included, and how you will present the written information.

9. Create your research report and share it with the class.

   Be sure to study carefully each other group's research reports.

   a) What did you learn from the other reports?

   b) How much agreement is there between the various reports on global change? Explain.

Teacher Commentary

Teaching Tip

Explain to your students that when scientists complete a research study and believe that they have found out something important, they share their findings with other scientists. This means that other scientists can review these findings, perhaps conduct their own research to verify these findings, or use them to ask further questions. One way in which scientists share their findings is by producing a report for other scientists to read. Sometimes these reports are published in scientific journals. Scientists also have other scientists visit them, or go to meetings where they can report and discuss their findings.

8. Students will need to decide within their groups how to organize different sections of the report. Make sure that everyone in the group is contributing to the completion of the report.

9. Students should understand that the reporting of findings is a critical part of scientific research. Encourage them to construct a presentation format that adequately covers the results of their scientific study.

Assessment Tool

Student Presentation Evaluation Form
Use the **Student Presentation Evaluation Form** as a simple guideline for assessing presentations. Adapt and modify the evaluation form to suit your needs. Provide the form to your students and discuss the assessment criteria before they begin their work.

Teaching Tips

Investigation 8 is purposely "open ended" and not prescriptive. Your students have a number of choices. The key assessment issue here is the quality of those choices, especially the level to which they are informed by the evidence and explanations they have derived from earlier investigations.

It is important that you allot class time after the presentations for students to discuss what impact their research has had on them personally. Help them to explore how the knowledge they now have has changed their views about climate and weather, and how they might change their behavior because of it.

Investigation 8: Climate Change Today

Digging Deeper

GLOBAL CLIMATE CHANGE

The Earth seems to be getting warmer. Graph 1 shows how yearly average temperature has changed since 1880. This is about the time when temperatures first began to be recorded in an organized way at weather stations around the world. The curve on the graph has lots of ups and downs, but there is a clear upward trend. Graphs like this have some uncertainties, however. For example, there is a problem about how to adjust the curve to take into account the growth of large cities, where so many of the weather stations are located. Urbanization causes temperature readings in large cities to be somewhat higher than in surrounding rural areas. The upward trend shown is almost certainly real, though. Other evidence, like the general shrinking of glaciers all around the world and the thinning of the Arctic ice pack, tell the same story.

As You Read...
Think about:
1. How has yearly average temperature changed since 1880?
2. How has global average temperature changed during the past 2400 years?
3. What is Earth's principal greenhouse gas?
4. Why has carbon dioxide increased over the past decades?
5. Why are computer models limited in their predictions of Earth's future climate?

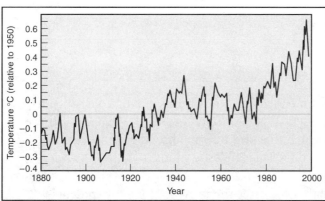

Graph 1

Graph 2 is similar to Graph 1. It shows global average temperature for the past 2400 years. The curve is less certain than the one in Graph 1, because it is based on various proxies for temperature. You learned

Teacher Commentary

Digging Deeper

This section provides text, graphs, and a photo that give students greater insight into the topic of global climate change. You may wish to assign the **As You Read** questions as homework to help students focus on the major ideas in the text.

As You Read...

1. Yearly average temperature has increased overall since 1880.

2. Global average temperature has experienced several "spikes" of higher and lower temperatures, but overall it has not varied greatly.

3. As stated in **Digging Deeper**, water vapor is the principal greenhouse gas. Carbon dioxide, also an important greenhouse gas, has increased sharply in recent decades as a result of burning fossil fuels and clearing tropical rain forests for agriculture.

4. The burning of fossil fuels has released carbon dioxide into the atmosphere.

5. Computer models are limited in their predictions of Earth's future climate because it is very difficult to incorporate some of the most important influences on climate, like the behavior of clouds.

Assessment Opportunity

You may wish to rephrase selected questions from the **As You Read** section into multiple choice or "true/false" format to use as a quiz. Use this quiz to assess student understanding and as a motivational tool to ensure that students complete the reading assignment and comprehend the main ideas.

About the Photo

Graph 1 illustrates recent trends in global average annual temperature relative to the average temperature in 1950. The data used to compile this graph are based on historical observations from weather stations around the world.

INVESTIGATING CLIMATE AND WEATHER

about these in the previous investigation. The highest temperatures are about 1°C above 20th century mean temperatures. The lowest temperatures are about 1°C below 20th century mean temperatures.

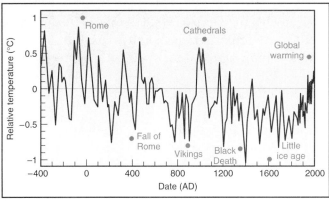

Graph 2

Here's a big question, and an important one. Has the increase in temperature since the beginning of the 20th century been caused by human activity, or is it just another natural upward "spike" like several during the past two millennia, shown in Graph 2? Most climatologists think that the upward trend in temperature during the 20th century is at least partly caused by human activity.

Several gases in the Earth's atmosphere are called greenhouse gases. The most important is water vapor, but carbon dioxide is important also. Early on in Earth's history, carbon dioxide was the principal greenhouse gas. Carbon dioxide has always been part of the Earth's atmosphere, but has been increasing more and more rapidly in recent times (Graph 3). Coal, oil, and natural gas are called fossil fuels, because they come from plant and animal material that was buried in the Earth's sediments. When they are burned, they add carbon dioxide gas to the atmosphere. Along with the other greenhouse gases, carbon dioxide absorbs some of the heat that the Earth's surface sends

Teacher Commentary

About the Photo

Graph 2 extends the data shown in Graph 1 back to 2400 years before present. This curve is based on interpretations of global temperature based on proxies to temperature like the isotopic composition of oxygen and hydrogen trapped in glacier ice.

Investigation 8: Climate Change Today

out to space and radiates heat back to the surface. That increases the Earth's average surface temperature. The effect is the same as with the glass of a greenhouse, although the process itself is not exactly the same.

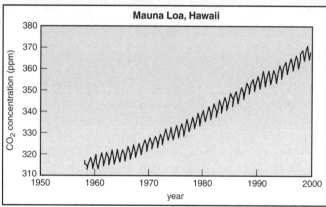

Graph 3

Several groups of climatologists have been developing computer models of the Earth's climate. They try to build in all of the important controls of climate into the model. Then they start the model with the present climate and let it run to see what the future climate will be. The models are not perfect, because it is very difficult to simulate some of the most important influences on climate. The behavior of clouds is especially tricky to model. The models have one thing in common, though. They predict that the Earth's surface temperature is likely to increase by as much as 2°C between 2000 and 2050 if the upward trend in

Teacher Commentary

About the Photo

Graph 3 shows the record of change in the levels of atmospheric carbon dioxide measured on top of Mauna Loa volcano in Hawaii. The high elevation and remote location of Mauna Loa make it an excellent indicator of the levels of atmospheric carbon dioxide away from any major industrial centers.

About the Photo

The photo on page C85 of the student text shows one of a number of mountain ranges that bound the Los Angeles Basin in southern California. Because of regional weather patterns, this basin often experiences temperature inversions whereby the air in the basin is slightly cooler than overlying air. That inhibits vertical mixing of the atmosphere and allows buildup of atmospheric pollutants, sometimes to dangerous levels. Strict control of emissions both from stationary sources and from motor vehicles has greatly improved air quality in the Los Angeles Basin in recent years.

INVESTIGATING CLIMATE AND WEATHER

atmospheric carbon dioxide continues. A look back at Graph 2 shows that a rise of 2°C would make the Earth's temperature much higher than during even the warmest periods in human history.

If the predictions about global warming come true, many things about the Earth's climate, aside from just temperature, are likely to change. Some regions will get more rainfall, and other regions will get less. The frequency and intensity of severe storms are likely to increase. As the world's glaciers continue to melt, sea level around the world will rise, by as much as half a meter or so. That might not sound like a lot to you, but think of the flooding that it would cause in coastal cities and islands around the world!

Will the predictions of the computer models come true? That is likely, although not certain. All that science can do is try to make likely predictions. How to act upon the predictions is for human society to decide.

Review and Reflect

Reflect

1. How did your ideas about how global climate will change in the future differ from those of other groups? Explain why there might be differences of opinion.

Thinking about the Earth System

2. Add any new connections that you have found between climate change and the Earth system (biosphere, atmosphere, hydrosphere, and geosphere) to your *Earth System Connection* sheet.

Thinking about Scientific Inquiry

3. What evidence did you find for future climate change?

Teacher Commentary

Review and Reflect

Reflect
1. Answers will vary.

Thinking about the Earth System
2. Answers will vary.

Thinking about Scientific Inquiry
3. Answers will vary.

> **Assessment Tool**
> Review and Reflect Journal–Entry Evaluation Sheet
> Use the general criteria on this evaluation sheet for assessing content and thoroughness of student work. Adapt and modify the sheet to meet your needs. Consider involving students in selecting and modifying the criteria for evaluating their reflections on **Investigation 8**.

Teacher Review

Use this section to reflect on and review the investigation. Keep in mind that your notes here are likely to be especially helpful when you teach this investigation again. Questions listed here are examples only.

Student Achievement

What evidence do you have that all students have met the science content objectives?

_____

Are there any students who need more help in reaching these objectives? If so, how can you provide this? _____

_____

What evidence do you have that all students have demonstrated their understanding of the inquiry processes? _____

_____

Which of these inquiry objectives do your students need to improve upon in future investigations? _____

_____

What evidence do the journal entries contain about what your students learned from this investigation? _____

_____

Planning

How well did this investigation fit into your class time? _____

_____

What changes can you make to improve your planning next time? _____

_____

Guiding and Facilitating Learning

How well did you focus and support inquiry while interacting with students?

_____

What changes can you make to improve classroom management for the next investigation or the next time you teach this investigation? _____

_____

Teacher Commentary

How successful were you in encouraging all students to participate fully in science learning? _____

_____

How did you encourage and model the skills values, and attitudes of scientific inquiry? _____

_____

How did you nurture collaboration among students? _____

_____

Materials and Resources

What challenges did you encounter obtaining or using materials and/or resources needed for the activity? _____

_____

What changes can you make to better obtain and better manage materials and resources next time? _____

_____

Student Evaluation

Describe how you evaluated student progress. What worked well? What needs to be improved? _____

_____

How will you adapt your evaluation methods for next time? _____

_____

Describe how you guided students in self-assessment. _____

_____

Self Evaluation

How would you rate your teaching of this investigation? _____

_____

What advice would you give to a colleague who is planning to teach this investigation? _____

_____

NOTES

Teacher Commentary

NOTES

Reflecting

Back to the Beginning
You have been investigating climate and weather in many ways. How have your ideas changed since the beginning of the investigation? Look at the following questions and write down your ideas in your journal.

- What information is contained in a weather report and how is this information obtained?
- What is the difference between climate and weather?
- What evidence is there that climate change has happened?
- What evidence is there that climate change is happening?

Thinking about the Earth System
At the end of each investigation, you thought about how your findings connected with the Earth system. Consider what you have learned about the Earth system. Refer to the *Earth System Connection* sheet that you have been building up throughout this module.

What connections between climate and weather and the Earth system have you been able to find?

Thinking about Scientific Inquiry
You have used inquiry processes throughout the module. Review the investigations you have done and the inquiry processes you have used.

- What scientific inquiry processes did you use?
- How did scientific inquiry processes help you learn about climate and weather?

Not so much an ending as a new beginning!

This investigation into climate and weather is now completed. However, this is not the end of the story. You will see the importance of climate and weather where you live, and everywhere you travel. Be alert for opportunities to observe the importance of climate and weather, and add to your understanding.

Reflecting

This is the point at which your students review what they have learned throughout the module. This review is very important. Allow students time to work on this in a thoughtful way.

Back to the Beginning

These four questions were used as a pre-assessment. Encourage students to complete this final review without looking at their journal entries from the beginning of the module. Their initial entries may influence their responses.

When students have completed their writing, encourage them to revisit their initial answers from the pre-assessment. Compare their writings at the end of the unit to their writings at the beginning. It is important that students are not left with the impression that they now know all there is to know about climate and weather. Emphasize that learning is a continuous process throughout our lives.

> ### Assessment Opportunity
>
> Comparisons between students' initial answers to these questions (in the pre-assessment at the beginning of the module) and those they are now able to give provide valuable data for assessment.

Thinking about the Earth System

Now that your students are at the end of this module, ask them to make connections between climate and weather and the different aspects of the Earth System. You may want to do this through a concept map. This is an opportunity for you to gauge how well students have developed their understanding of the Earth System for assessment purposes.

Thinking about Scientific Inquiry

To help students understand the relevance of these processes to their lives, ask them to think of everyday examples of when they use these processes (finding out where a misplaced book has gone; forming an opinion about a new TV show; winning an argument).

NOTES

Appendices

Investigating Climate and Weather
Alternative End-of-Module Assessment

Part A: Matching.
Write the letter of the term from Column B that best matches the description in Column A.

Column A	Column B
1. Line on a weather map connecting points of equal air pressure.	A. Air temperature
2. Presentation of recent weather events.	B. Anemometer
3. A gas in the air that has an important effect upon weather.	C. Barometer
	D. Forecast
4. Instrument used to measure wind speed.	E. Hydrometer
5. Line on a weather map connecting points of equal air temperature.	F. Isobar
	G. Isometer
6. Instrument used to measure air pressure.	H. Isotherm
7. Prediction of upcoming weather.	I. Nitrogen
	J. Water vapor
	K. Weather report
	L. Wind vane

Part B: Multiple Choice.
Provide the letter of the choice that best answers the questions or completes the statement.

8. Clouds are most likely to form when
   A. Cold fronts make warm dry air descend mountain slopes.
   B. Cold dry air is made to rise.
   C. Warm humid air is made to rise.
   D. Warm dry air is made to rise.

9. Winds tend to blow:
   A. Inward toward the center of a zone of low pressure.
   B. Inward toward the center of a zone of high pressure.
   C. Parallel to isotherms.
   D. Perpendicular to isotherms.

10. Over which region is a warm, humid air mass most likely to form?
    A. Central Canada.
    B. The Gulf of Mexico.
    C. The Rocky Mountains.
    D. The northwest coast of the United States.

11. Which of the following statements about air pressure is true?
    A. Air pressure increases with altitude.
    B. Air pressure decreases with altitude.
    C. Air pressure decreases from west to east.
    D. Air pressure does not change with altitude.

12. Which of the following describes how weather changes when a cold front moves into an area?
    A. Temperatures fall, and the sky remains clear.
    B. Temperatures fall, and thunderstorms are likely to form.
    C. Temperatures remain constant, and the sky remains clear.
    D. Temperatures rise, and thunderstorms are likely to form.

13. What is the usual relationship between air temperature and altitude in the lowest layer of the atmosphere?
    A. Temperature does not change with altitude.
    B. Temperature increases with altitude.
    C. Temperature decreases with altitude.
    D. Temperature decreases with altitude, except at the equator or the poles.

14. If a meteorologist announces that a high-pressure system is moving into your area, which of the following weather conditions could you accurately predict?
    A. Clear blue skies and nice weather for at least a day.
    B. Warm, wet stormy weather for the next six hours.
    C. Warm, wet stormy weather for at least one day.
    D. Cold, wet stormy weather for at least one day.

15. What would happen when the relative humidity reaches 100%?
    A. Liquid water would no longer evaporate.
    B. Liquid water would evaporate more rapidly.
    C. Water vapor would condense.
    D. Water vapor would no longer condense.

16. Which statement best describes the relationship between weather and climate?
    A. Weather is a general prediction of future climate.
    B. Weather and climate are unrelated.
    C. Climate is an accurate method of predicting how weather will change each week.
    D. Climate is a long-term average of weather.

17. How does the climate of a coastal city where wind blows from the ocean compare to a city far inland (assuming the cities are at the same latitude)?
    A. The coastal city would have a larger range of daily temperatures.
    B. The coastal city would have a smaller range of daily temperatures.
    C. The coastal city would be warmer.
    D. The coastal city would be colder.

18. Studies of the deposits of sediments left by glaciers have allowed scientists to conclude that:
    A. Earth was once completely covered by ice.
    B. The Earth's climate was once cold, and now it is warm.
    C. Earth has experienced many ice ages and warming periods.
    D. The Earth's average climate will keep warming until all ice has melted.

19. In order for a cloud to form, which of the following must be present in air?
    A. High pressure and relative humidity.
    B. Oxygen and hydrogen.
    C. Water vapor and condensation nuclei.
    D. Nitrogen, oxygen, and carbon dioxide.

20. The process in which liquid water changes into water vapor is called:
    A. Freezing
    B. Melting
    C. Evaporation
    D. Condensation

21. Freezing of water on the Fahrenheit scale occurs at:
    A. 0°
    B. 32°
    C. 100°
    D. 212°

22. A wind that moves form north to south is called a:
    A. Polar wind
    B. Arctic wind
    C. North wind
    D. South wind

Part C: Essay

23. How does a thermometer work?

24. In your response, explain how vegetation is an indicator of climate. Compare a tropical rainforest with a desert.

Answer Key

1. F
2. K
3. J
4. B
5. H
6. C
7. D
8. C
9. A
10. B
11. B
12. B
13. C
14. A
15. C
16. D
17. B
18. C
19. C
20. D
21. B
22. C

23. Thermometers have a small glass bulb at the end of a long glass tube that is sealed at both ends. A very small tubular hollow center, called the bore, extends from the bulb to the opposite end of the tube. The tube is graduated in degrees Fahrenheit or degrees Celsius, or both. A liquid in the bulb—almost always either mercury or colored alcohol—can move up the bore. When the air temperature rises, the liquid in the bulb expands more than the glass of the thermometer, causing the liquid to rise in the tube. Because the bore is very narrow, a small temperature change causes a large change in the length of the liquid in the tube.

24. Tropical rainforests are areas with moderate to high temperatures and abundant rainfall throughout the year and are heavily forested. Deserts are regions with not much rainfall and scarce vegetation.

Investigating Earth Systems Assessment Tools

Assessing the Student *IES* Journal

- Journal–Entry Evaluation Sheet
- Journal–Entry Checklist
- Key–Question Evaluation Sheet
- Investigation Journal–Entry Evaluation Sheet
- Review and Reflect Journal–Entry Evaluation Sheet

Assisting Students with Self Evaluation

- Group–Participation Evaluation Sheet I
- Group–Participation Evaluation Sheet II

Assessing the Final Investigation

- Final Investigation Evaluation Sheet
- Student Presentation Evaluation Form

References

- Doran, R., Chan, F., and Tamir, P. (1998). *Science Educator's Guide to Assessment*.
- Leonard, W.H., and Penick, J.E. (1998). *Biology – A Community Context*. South-Western Educational Publishing. Cincinnati, Ohio.

Journal–Entry Evaluation Sheet

Name: _____ Date: _____ Module: _____

Explanation: The journal is an important component of each *IES* module. In using the journal as you investigate Earth science questions, you are mirroring what scientists do. The criteria, along with others that your teacher may add, will be used to evaluate the quality of your journal entries. Use these criteria, along with instructions within investigations, as a guide.

Criteria

1. Entry Made
      1   2   3   4   5   6   7   8   9   10   _____
   Blank   Nominal   Above average Thorough

2. Detail
      1   2   3   4   5   6   7   8   9   10   _____
   Few dates   Half the time Most days   Daily
   Little detail Some detail   Good detail Excellent detail

3. Clarity
      1   2   3   4   5   6   7   8   9   10   _____
   Vague   Becoming clearer Clearly expressed
   Disorganized   well organized

4. Data Collection/Analysis
      1   2   3   4   5   6   7   8   9   10   _____
   Data collected   Data collected,   Data collected
   Not analyzed   some analyzed   and analyzed

5. Originality
      1   2   3   4   5   6   7   8   9   10   _____
   Little evidence   Some evidence   Strong evidence
   of originaiity   of originality   of originality

6. Reasoning/Higher-Order Thinking
      1   2   3   4   5   6   7   8   9   10   _____
   Little evidence   Some evidence   Strong evidence
   of thoughtfulness of thoughtfulness of thoughtfulness

7. Other
      1   2   3   4   5   6   7   8   9   10   _____

8. Other
      1   2   3   4   5   6   7   8   9   10   _____

Journal–Entry Checklist

Name: _____ Date: _____ Module: _____

Explanation: The journal is an important component of each *IES* module. In using the journal as you investigate Earth science questions, you are mirroring what scientists do. The criteria, along with others that your teacher may add, will be used to evaluate the quality of your journal entries. Use these criteria, along with instructions within investigations, as a guide.

Criteria

1. Makes entries _____

2. Provides dates and details _____

3. Entry is clear and organized _____

4. Shows data collected _____

5. Analyzes data collected _____

6. Shows originality in presentation _____

7. Shows evidence of higher-order thinking _____

8. Other _____

9. Other _____

Total Earned _____

Total Possible _____

Comments:

Key–Question Evaluation Sheet

Name: _____ Date: _____ Module: _____

	No Entry		Fair		Strong
Shows evidence of prior knowledge	0	1	2	3	4
	No Entry		Fair		Strong
Reflects discussion with classmates	0	1	2	3	4

Additional Comments

Key–Question Evaluation Sheet

Name: _____ Date: _____ Module: _____

	No Entry		Fair		Strong
Shows evidence of prior knowledge	0	1	2	3	4
	No Entry		Fair		Strong
Reflects discussion with classmates	0	1	2	3	4

Additional Comments

Key–Question Evaluation Sheet

Name: _____ Date: _____ Module: _____

	No Entry		Fair		Strong
Shows evidence of prior knowledge	0	1	2	3	4
	No Entry		Fair		Strong
Reflects discussion with classmates	0	1	2	3	4

Additional Comments

Investigation Journal–Entry Evaluation Sheet

Name: _____ Date: _____ Module: _____

Criteria

1. Completeness of written investigation
       1 2 3 4 5 6 7 8 9 10   _____
   Blank   Incomplete   Thorough

2. Participation in investigations
       1 2 3 4 5 6 7 8 9 10   _____
   None or little;   Needs minimal guidance, Leads, is inquisitive,
   unable to guide sometimes helping others persistent, focused
   self

3. Skills attained
       1 2 3 4 5 6 7 8 9 10   _____
   Few skills   Tends to use some   High degree of
   evident   appropriate skills   appropriate skills used

4. Investigation Design
       1 2 3 4 5 6 7 8 9 10   _____
   Variables not   Sometimes   Considers variables
   considered   considers variables,   Sound rationale for
   techniques   uses logical techniques techniques
   illogical

5. Conceptual understanding of content
       1 2 3 4 5 6 7 8 9 10   _____
   No evidence   Approaches understanding Exceeds expectations
   of understanding of most concepts   for content attainment

6. Ability to explain/discuss inquiry
       1 2 3 4 5 6 7 8 9 10   _____
   Unable to   Some ability to   Uses scientific reasoning
   articulate   explain/discuss   to explain any
   scientific thought the inquiry   aspect of the inquiry

7. Other
       1 2 3 4 5 6 7 8 9 10   _____

8. Other
       1 2 3 4 5 6 7 8 9 10   _____

Review and Reflect Journal–Entry Evaluation Sheet

Name: _____ Date: _____ Module: _____

Criteria	Blank		Fair		Excellent	
Thoroughness of answers	0	1	2	3	4	5
Content of answers	0	1	2	3	4	5
Other	0	1	2	3	4	5

Review and Reflect Journal–Entry Evaluation Sheet

Name: _____ Date: _____ Module: _____

Criteria	Blank		Fair		Excellent	
Thoroughness of answers	0	1	2	3	4	5
Content of answers	0	1	2	3	4	5
Other	0	1	2	3	4	5

Review and Reflect Journal–Entry Evaluation Sheet

Name: _____ Date: _____ Module: _____

Criteria	Blank		Fair		Excellent	
Thoroughness of answers	0	1	2	3	4	5
Content of answers	0	1	2	3	4	5
Other	0	1	2	3	4	5

Group–Participation Evaluation Sheet I

Key:
4 = Worked on his/her part and assisted others
3 = Worked on his/her part
2 = Worked on part less than half the time
1 = Interfered with the work of others
0 = No work

My name is _____ .   I give myself a _____

The other people in my group are: I give each person:

A. _____   _____

B. _____   _____

C. _____   _____

D. _____   _____

Key:
4 = Worked on his/her part and assisted others
3 = Worked on his/her part
2 = Worked on part less than half the time
1 = Interfered with the work of others
0 = No work

My name is _____ .

The other people in my group are:

A. _____

B. _____

C. _____

D. _____

Group–Participation Evaluation Sheet II

Name: _____ Date: _____ Module: _____

Key:
Highest rating _____
Lowest rating _____

1. In the chart, rate each person in your group, including yourself.

	Names of Group Members				
Quality of Work					
Quantity of Work					
Cooperativeness					
Other Comments _____					

2. What went well in your investigation?

3. If you could repeat the investigation, how would you change it?

Investigating Climate and Weather 333

Final Investigation Evaluation Sheet

Alerting students
Before your students begin the final investigation, they must understand what is expected of them and how they will be evaluated on their performance. Review the task thoroughly, setting time guidelines and parameters (whom they may work with, what materials they can use, etc.). Spell out the evaluation criteria for each level of proficiency shown below. Use three categories for a 3-point scale (Achieved, Approaching, Attempting). If you prefer a 5-point scale, add the final two categories.

Name: _____ Date: _____ Module: _____

	Understanding of concepts and inquiry	Use of evidence to explain and support results	Communication of ideas	Thoroughness of work
Exceeding proficiency 5	Demonstrates complete and unambiguous understanding of the problem and inquiry processes used.	Uses all evidence from inquiry that is factually relevant, accurate, and consistent with explanations offered.	Communicates ideas clearly and in a compelling and elegant manner to the intended audience.	Goes beyond all deliverables agreed upon for the project and has extended the data collection and analysis.
Achieved proficiency 4	Demonstrates fairly complete and reasonably clear understanding of the problem and inquiry processes used.	Uses the major evidence from inquiry that is relevant and consistent with explanations offered.	Communicates ideas clearly and coherently to the intended audience.	Includes all of the deliverables agreed upon for the project.
Approaching proficiency 3	Demonstrates general, yet somewhat limited understanding of the problem and inquiry processes used.	Uses evidence from inquiry to support explanations but may mix fact with opinion, omit significant evidence, or use evidence that is not totally accurate.	Completes the task satisfactorily but communication of ideas is incomplete, muddled, or unclear.	Work largely complete but missing one of the deliverables agreed upon for the project.
Attempting proficiency 2	Demonstrates only a very general understanding of the problem and inquiry processes used.	Uses generalities or opinion more than evidence from inquiry to support explanations.	Communication of ideas is difficult to understand or unclear.	Work missing several of the deliverables agreed upon for the project.
Non-proficient 1	Demonstrates vague or little understanding of the problem and inquiry processes used.	Uses limited evidence to support explanations or does not attempt to support explanations.	Communication of ideas is brief, vague, and/or not understandable.	Work largely incomplete; missing many of the deliverables agreed upon for the project.

Student Presentation Evaluation Form

Student Name_____ Date_____

Topic_____

	Excellent		Fair		Poor
Quality of ideas	4	3	2	1
Ability to answer questions	4	3	2	1
Overall comprehension	4	3	2	1

COMMENTS
_____
_____
_____
_____

Student Presentation Evaluation Form

Student Name_____ Date_____

Topic_____

	Excellent		Fair		Poor
Quality of ideas	4	3	2	1
Ability to answer questions	4	3	2	1
Overall comprehension	4	3	2	1

COMMENTS
_____
_____
_____
_____

Blackline Master *Climate and Weather* P.1

Questions about Climate and Weather

- What information is contained in a weather report and how is this information obtained?

- What is the difference between climate and weather?

- What evidence is there that climate change has happened?

- What evidence is there that climate change is happening?

Use with *Climate and Weather* Pre-assessment.

Blackline Master *Climate and Weather* P.2

Student Journal Cover Sheet
Investigating Climate and Weather

Name: _____

Group Members:

1. _____

2. _____

3. _____

4. _____

Teacher: _____

Class: _____

Dates of Investigation:

Start _____ Complete _____

Keep this journal with you at all times during your study of
Investigating Climate and Weather

Use with *Climate and Weather* Pre-assessment.

Blackline Master *Climate and Weather* **I.1**

Name: _____

Earth System Connection Sheet

When you finish an investigation, use this sheet to record any links you can make with the Earth system. By the end of the module you should have as complete a diagram as possible.

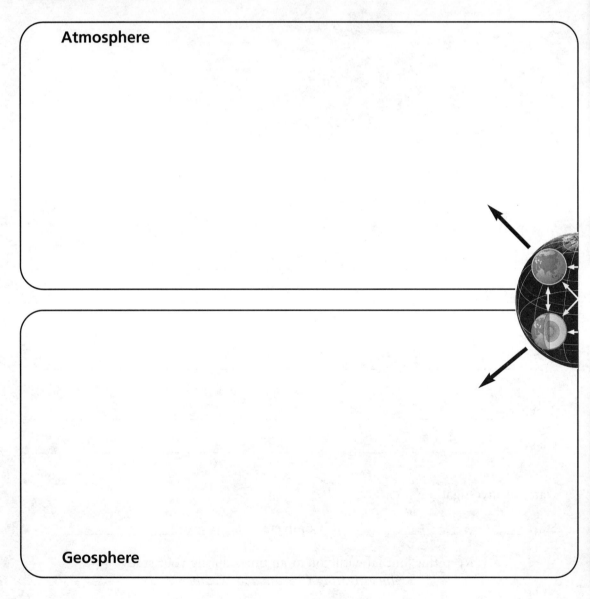

Atmosphere

Geosphere

Use with *Climate and Weather* Introducing the Earth System.

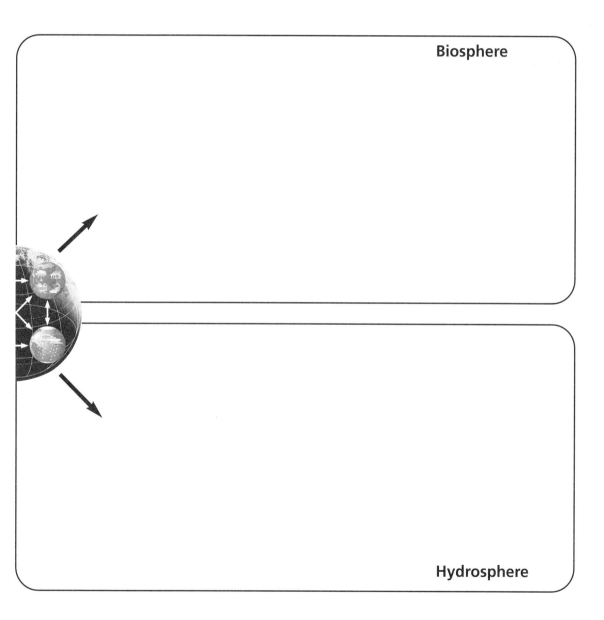

Blackline Master *Climate and Weather* **P.2**

Inquiry Processes

 • Explore questions to answer by inquiry.

 • Design an investigation.

 • Conduct an investigation.

 • Collect and review data using tools.

 • Use evidence to develop ideas.

 • Consider evidence for explanations.

 • Seek alternative explanations.

 • Show evidence and reasons to others.

 • Use mathematics for science inquiry.

Use with *Climate and Weather* Inquiry Processes.

Blackline Master *Climate and Weather* 1.1

Making Weather Observations
Temperature

Temperature is measured with a thermometer. Thermometers are usually glass tubes filled with colored alcohol. As air temperature increases, the level of the colored alcohol in the tube rises and, as air temperature decreases, the level falls. Mount your thermometer in a shady place one to two meters above the ground. Make sure that it will not be blown away be a strong wind. Also make sure that the thermometer is not directly in the Sun. If you decide not to leave the thermometer outside the whole time, give it a few minutes to reach the outdoor temperature after taking it outside. Read the thermometer with your eyes as level with the thermometer as possible. Record temperatures in degrees Fahrenheit and degrees Celsius to the nearest integer.

Use with *Climate and Weather* Investigation 1: Observing Weather

Blackline Master *Climate and Weather* 1.2

Making Weather Observations
Wind Direction

Wind direction is determined by noting the direction that the wind is blowing from, not the direction that the wind is blowing to. Wind direction is recorded as north, northeast, east, southeast, south, southwest, west, or northwest. Wind direction can be estimated by noting which way a flag is blowing, or you can construct a simple wind vane to help you determine wind direction. One example of how to construct a wind vane is described below.

You will need the following materials:
- tag board or manila file folder
- straight pin
- scissors
- glue
- sharpened pencil with a new eraser
- plastic drinking straw
- modeling clay
- paper plate, with the compass directions north, south, east, and west labeled

Cut an arrow point about 5 cm long and an arrow tail about 7 cm long out of the tag board or manila file folder. Make 1-cm cuts at the ends of the straw and slide the arrow point and the arrow tail into the cuts. Push the straight pin through the middle of the straw and into the eraser end of the pencil. Stick the sharp end of the pencil into a lump of modeling clay. The plate will be the base of your wind vane. Put the clay on a paper plate. Blow on the wind vane and make sure that the arrow can spin freely.

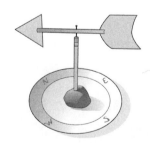

Take your wind vane to a location that is free from wind obstructions and high off the ground (but remember that you must be able to see the plate to record the wind direction). Secure your wind vane to make sure that it will not blow over. Use the North Star to line up the "North" marker on the paper plate in order to properly position the directional letters. You can also use a pocket compass to line up the directional letters, but remember that the readings you take will be relative to magnetic north, rather than true north.

The arrowhead will point into the wind and therefore will indicate the direction from which the wind is blowing. The arrow may vary in position from minute to minute, so you will want to observe the arrow for several minutes and determine the dominant wind direction. Also, try to make your observations at the same time each day.

Use with *Climate and Weather* Investigation 1: Observing Weather

Blackline Master *Climate and Weather* 1.3

Making Weather Observations
Wind Speed

Estimates of wind speed can be made by noting the effect of wind on branches of trees, leaves, smoke, etc., and relating these effects to the Beaufort Wind Scale, which is shown on page C5 of your text. Wind speed can be measured using an instrument called an anemometer. Instructions for constructing a simple anemometer are given below.

You will need the following materials:
- 5 paper cups
- 3 wooden dowels, about 10 inches long each
- modeling clay
- piece of wood, approximately 10 inches x 10 inches, to serve as a base

Set up the paper cups and dowels as shown in the illustration below. Place an "X" on one of the outer cups. Place modeling clay on the wooden base, and place the center dowel in the clay. Make sure that the anemometer will not tip over.

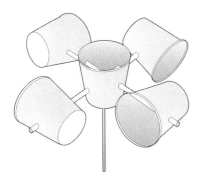

To calibrate the anemometer, have someone drive you in a car on a windless day. Hold the instrument out the window at arm's length and have the driver proceed at a constant speed starting at 5 mph. Count the number of turns the cups make in 60 s. You can use the cup marked with the "X" as a reference point. Repeat your counting at 10 mph, 15 mph, and so on to 35 mph. Produce a data table with wind speed (how fast the car was moving) in one column and the number of cup revolutions per minute in the other. You can then use the data table to translate revolutions per minute into wind speed in miles per hour.

You can also determine the wind speed by first counting the number of revolutions made by the anemometer per minute. Next, find the circumference of the circle (in feet) made by the rotating paper cups. Multiply the revolutions per minute by the circumference of the circle (in feet per revolution), and you will have the wind speed in feet per minute (you may want to convert this to mph, to see where it fits on the Beaufort Wind Scale). This will be an underestimate, because the cups do not move quite as fast as the wind. That is because there is some wind resistance on the cups as they make the "return trip" back against the wind.

Set up your anemometer in a location free from wind obstructions (trees, buildings, etc.) and as high as possible, remembering that you must be able to count the number of cup revolutions.

Use with *Climate and Weather* Investigation 1: Observing Weather

Blackline Master *Climate and Weather* 1.4

Making Weather Observations Clouds

Cloud Types
There are four basic groups of clouds: high clouds, middle clouds, low clouds, and clouds with vertical growth. The names of different clouds are based on the Latin word that describes their appearance from the ground: *cirrus* means "curl of hair," *stratus* means "layer," *cumulus* means "heap," and *nimbus* means "rain." Prefixes are added to cloud names to give further description: "*cirr*" refers to high-level clouds and "*alto*" refers to middle-level clouds.

High clouds (cloud base above 6000 m, or 20,000 ft)
Cirrus: thin, wispy, and feathery; also called "mares' tails."
Cirrocumulus: puffy and patchy in appearance; may form wave-like patterns that are sometimes called "mackerel sky."
Cirrostratus: light gray or white; thin sheets of ice crystals that usually cover much of the sky; may create a halo around the Sun or moon.

Middle clouds (cloud base 2000–6000 m, or 6500–20,000 ft)
Altostratus: uniformly gray or bluish white layers; covers most of the sky in a thin sheet; Sun or moon may shine through as a bright spot, as if viewed through frosted glass.
Altocumulus: roll-like puffy, patchy clouds, grouped in large sheets; also called "sheep's back" clouds.

Low clouds (cloud base below 2000 m, or 6500 ft)
Stratus: light or dark gray low cloud; covers most of the sky; may produce drizzle.
Fog is a stratus cloud in contact with the ground.
Nimbostratus: thick dark gray cloud producing rain or snow; often with a ragged base.
Stratocumulus: irregular masses of clouds, often rolling or puffy in appearance.

Vertically developed clouds (cloud thickness from 500 to 18,000 m, or 1600 to 60,000 ft)
Cumulus: puffy white clouds resembling cotton balls, popcorn, or cauliflower heads floating in the sky; usually have almost flat bottoms; occur individually or in groups.
Cumulonimbus: thunderstorm clouds; tall, billowing towers of puffy clouds with flat bases; can have sharp, well-defined edges or anvil shape at the top; often produces rain and sometimes hail, strong winds, or tornadoes.

Observe which type or types of clouds are in the sky. If you are not sure about the types of clouds, write a few sentences in your notebook to describe how they look.

Cloud Cover
Cloud coverage is an estimate of how much of the sky is covered by clouds. It is classified into the following four categories.

Overcast: 90% or more of the sky is covered by clouds.
Broken: approximately 50% to 90% of the sky is covered by clouds ("mostly cloudy").
Scattered: approximately 10% to 50% of the sky is covered by clouds ("partly cloudy").
Clear: approximately 10% or less of the sky is covered by clouds.

Observe how much of the sky is covered by clouds and record this in your journal.

Use with *Climate and Weather* Investigation 1: Observing Weather

Blackline Master *Climate and Weather* 1.5

Making Weather Observations Precipitation

Precipitation
Precipitation is measured as the amount that falls over a given period of time (usually 24 hours). Therefore, you will want to record your measurements at approximately the same time each day. When making an observation while there is precipitation, note the time precipitation began and ended. Also, you will want to make note of what kind of precipitation it is (rain, snow, sleet, hail) and the intensity of precipitation (light, moderate, or heavy). To estimate precipitation intensity, note the distance to landmarks (like a tree or mailbox) and determine which landmarks are visible during the precipitation.

Rainfall (or melted snow, sleet, or hail) is measured using an instrument called a rain gauge. Instructions for making a rain gauge are given below.

You will need the following materials:
- tin can (like a coffee can) that is open at top, with no lip (so you can pour rain out of it easily)
- ruler
- tall glass jar or graduated cylinder
- block of wood, log, or box stuck in the ground on end to use as a platform

The can should sit on top of the platform on top so that it will not fall off. The open top of the can should be level (not at an angle) and about 30 cm above the ground. The tin can will serve as your rain collection device. To get an actual measurement of the amount of rain that fell, you will use the tall glass jar or graduated cylinder. If you are using a glass jar, it should have as small a diameter as possible, still allowing for it to hold enough water to fill the tin can to a depth of several centimeters. The jar should also be as cylindrical as possible (i.e. not tapering).

To calibrate the jar, fill the tin can with enough water to fill the jar. This will allow for maximum accuracy in your calibration. Measure the depth of water in the tin can. Adjust the water level in the tin can so that its depth is equal to the nearest integer centimeter (for example, if there are 4.2 centimeters of water in the can, pour off some of the water until there are exactly 4 centimeters of water in the can). This will make the calibration easier. Pour this water into the glass jar and mark the water level with a grease pencil or paint. From this first mark, use a ruler to divide up the part of the jar below the mark into a number of equal increments that is the same as the number of centimeters of water that you had in the tin can. (If you had four centimeters of water in the can, make 4 equal increments.) Each one of these marks will denote one centimeter of rainfall. Now, subdivide the space between each of the marks into ten equal increments. Repeat this process for inches as well. If you are using a graduated cylinder, use the volumetric markings to calibrate your rain gauge to both cm and inches of rainfall.

Place the tin can in as open an area as possible, far from buildings, trees, etc., which might interfere with your rain catch. Also make sure that the can will not blow over. After the

Use with *Climate and Weather* Investigation 1: Observing Weather

precipitation has ended, pour the water that has collected in the tin can into the calibrated jar and read off rainfall in centimeters and inches. Why is this more accurate than simply sticking a ruler inside the tin can? Read the rain gauge at least once or twice a day, at approximately the same time.

Snowfall can be measured by using a ruler to record snow depth. Measurements should be taken at several places in an area, avoiding areas where snow has been blown into a drift. Record all of your measurements and take the average of the readings.

Blackline Master *Climate and Weather* 1.6

The Beaufort Wind Scale

Beaufort Number	Kilometers per hour	Miles per hour	Wind Name	Land Indication
0	<1	<1	calm	smoke rises vertically
1	1–5	1–3	light air	smoke drifts
2	6–11	4–7	light breeze	leaves rustle
3	12–19	8–12	gentle breeze	small twigs move
4	20–29	13–18	moderate breeze	small branches move
5	30–38	19–24	fresh breeze	small trees sway
6	39–50	25–31	strong breeze	large branches move
7	51–61	32–38	moderate gale	whole trees move
8	62–74	39–46	fresh gale	twigs break off trees
9	75–86	47–54	strong gale	branches break
10	87–101	55–63	whole gale	some trees uprooted
11	102–119	64–73	storm	widespread damage
12	>120	>74	hurricane	severe destruction

Use with *Climate and Weather* Investigation 1: Observing Weather

Blackline Master *Climate and Weather* **1.7**

Types of Clouds

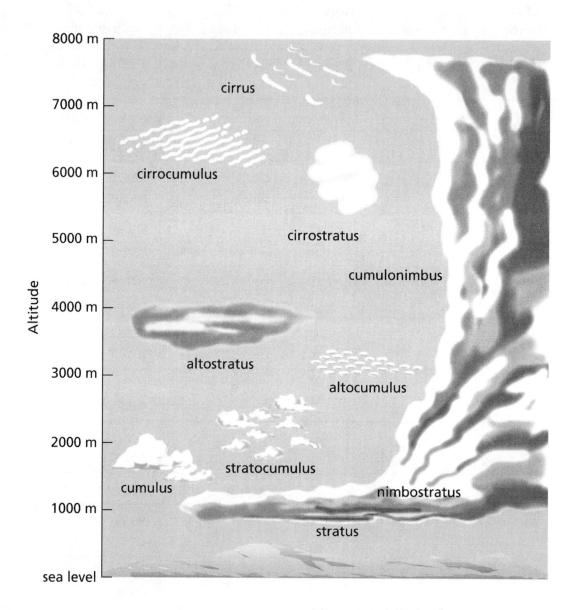

Use with *Climate and Weather* Investigation 1: Observing Weather

Blackline Master *Climate and Weather* 1.8

Standard Symbols for Weather Observation

Station model diagram:
- Air temperature (°C) → 18.5
- Visibility (× 100m) → 45
- Weather condition → ∞
- Dew point (°C) → 13.5
- Wind speed and direction
- Type of high cloud
- Type of middle cloud
- 101.2 → Barometric pressure (kPa)
- +0.4 → Change in barometric pressure (kPa)
- Cloud cover
- Type of low cloud

Weather Condition	Wind Speed (km/h)	Some Types of High Clouds	Some Types of Middle Clouds	Some Types of Low Clouds	Cloud Cover
• Rain	◉ Calm	⌒ cirrus	∠ altostratus	⌒ cumulus	○ Clear
∾ Freezing rain	— 3	2 cirrostratus	ω altocumulus	⌵ stratocumulus	⊖ Scattered clouds
* Snow	—| 9	∾ cirrocumulus		— stratus	◐ Partly cloudy
⍰ Thunderstorm	—|| 19			⌒ cumulonimbus	● Cloudy
≡ Fog	—||| 28				⊗ Sky obscured
∞ Haze	—|||| 37				
⌒ Dew	—||||| 45				
	⇒ 100				

Use with *Climate and Weather* Investigation 1: Observing Weather

Blackline Master *Climate and Weather* 2.1

Data Table for Weather Words

Weather Word	Descriptors Used	Definition	Instrument Used for Measuring	Unit of Measurement
Wind Speed				
Wind Direction				
Clouds				
Kind of Precipitation				
Amount of Precipitation				
Temperature				
Pressure				
Other				

Use with *Climate and Weather* Investigation 2: Comparing Weather Reports

Blackline Master *Climate and Weather* 2.2

Data Table for Kinds of Weather Information in Weather Reports

Weather Report Source	Date and Time of Report	Temperature	Wind Speed	Humidity	Cloud Cover	Other
Local Paper						
National Newspaper						
Radio						
Internet						
Television						
Phone						

Use with *Climate and Weather* Investigation 2: Comparing Weather Reports

Blackline Master *Climate and Weather* 2.3

Data Table for Weather Forecast for a Three-Day Period

Source	Day 1 Forecast	Day 1 Actual Data	Day 2 Forecast	Day 2 Actual Data	Day 3 Forecast	Day 3 Actual Data
Local Paper						
National Newspaper						
Radio						
Internet						
Television						
Phone						

Use with *Climate and Weather* Investigation 2: Comparing Weather Reports

Blackline Master *Climate and Weather* 3.1

Weather Map of the United States Showing Sky Conditions

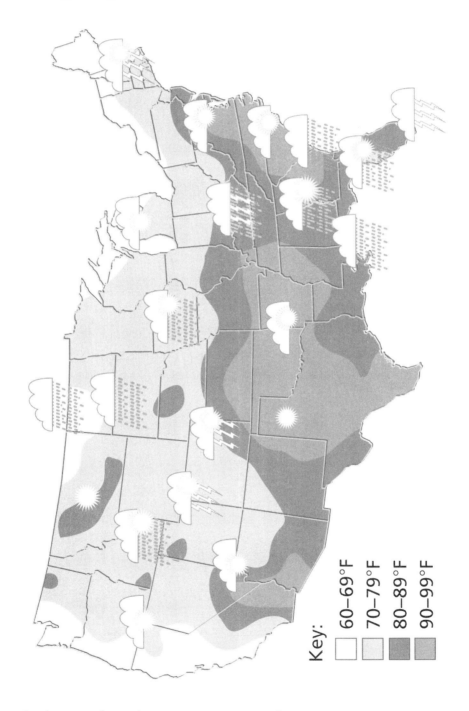

Use with *Climate and Weather* Investigation 3: Weather Maps

Blackline Master *Climate and Weather* 3.2

Weather Map of the United States Showing High- and Low-Pressure Systems

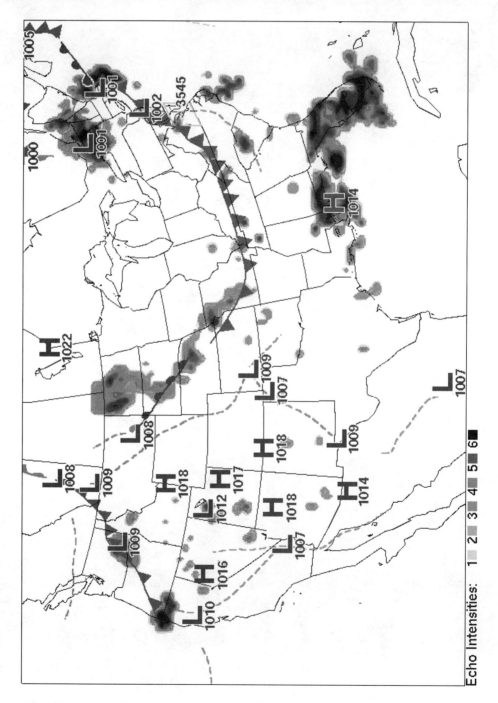

Use with *Climate and Weather* Investigation 3: Weather Maps

Blackline Master *Climate and Weather* 3.3

Weather Map of the United States Showing Temperatures

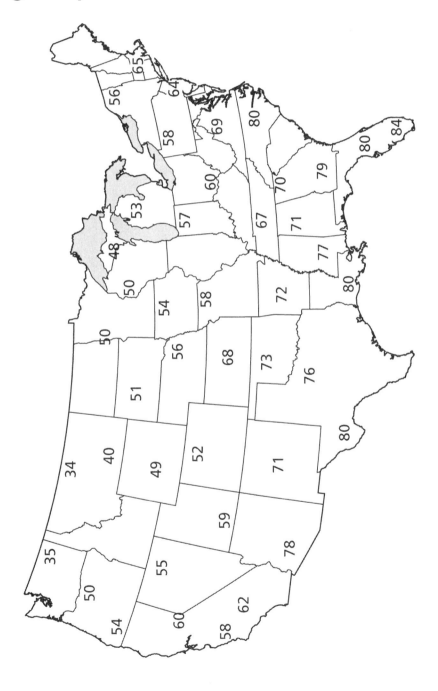

Use with *Climate and Weather* Investigation 3: Weather Maps

Blackline Master *Climate and Weather* **3.4**

Air Masses in North America

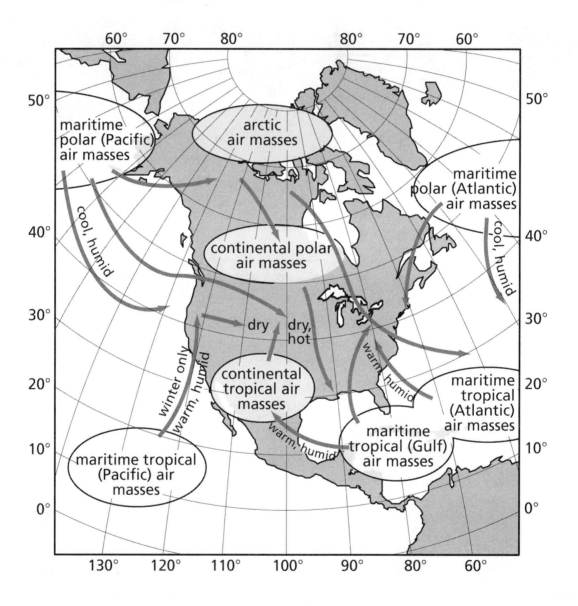

Use with *Climate and Weather* Investigation 3: Weather Maps

Blackline Master *Climate and Weather* 3.5

Warm Fronts and Cold Fronts

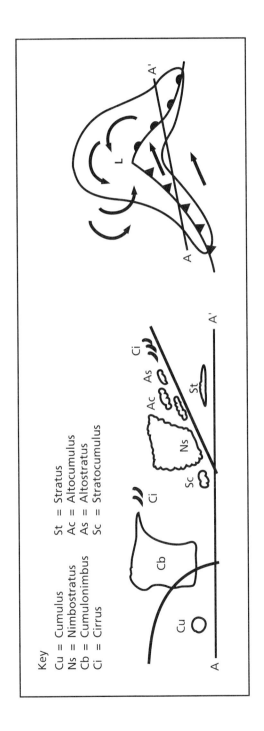
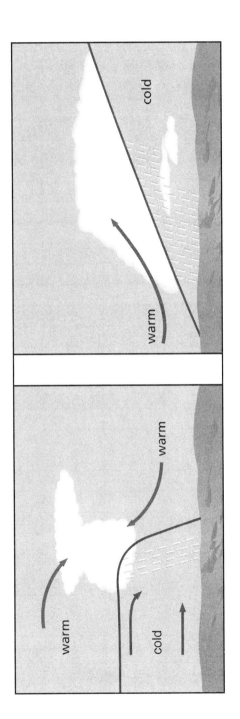

Use with *Climate and Weather* Investigation 3: Weather Maps

Blackline Master *Climate and Weather* 4.1

Radiosonde Data

| Radiosonde Data June 26, 2001 ||||
| Jacksonville, Florida || Fairbanks, Alaska ||
Temperature (°C)	Altitude (m)	Temperature (°C)	Altitude (m)
20.6	9	17	138
24.2	88	16.4	197
24.6	203	14.1	610
24.8	327	12.8	842
23.2	610	9.7	1219
21.6	884	7	1545
18.4	1219	3	2164
13.2	1829	−2.5	2923
11.2	2134	−6.4	3658
5	3224	−11.4	4540
−0.3	4267	−16.3	5180
−4.3	4997	−22.1	5970
−12.7	6096	−27.7	6746
−23.7	7570	−32.7	7620
−28.3	8230	−38.7	8469
−39.1	9610	−44.9	9300
−43.7	10830	−54.7	10668
−56.7	11983	−57.7	11900
−59.7	13766	−51.7	12719
−64.1	15240	−50.3	15240
−64.9	16540	−49.3	16619
−67.7	17692	−50.1	18601
−64.9	18700	−48.1	20950
−59.7	20780	−48.6	22555
−54.2	22555	−46.6	23774
−49.8	26518	−43.6	25603
−50.7	27432	−40	27432
−42.1	30480	−36.3	29403
−39.9	31270	−33.3	31840
−39	33223	−28.1	33528

Use with *Climate and Weather* Investigation 4: Radiosondes, Satellites, and Radar

Blackline Master *Climate and Weather* **4.2**

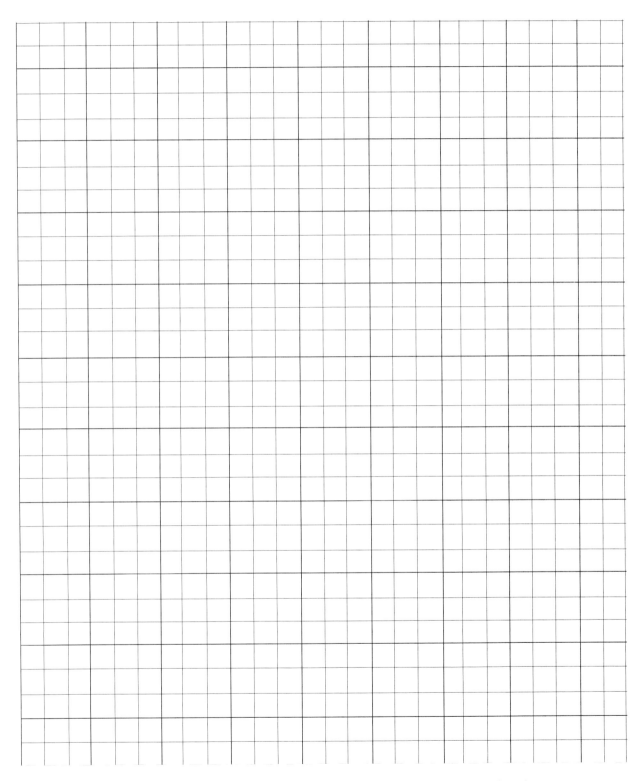

Use with *Climate and Weather* Investigation 4: Radiosondes, Satellites, and Radar

Blackline Master *Climate and Weather* 4.3

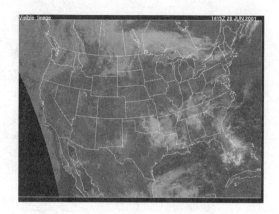 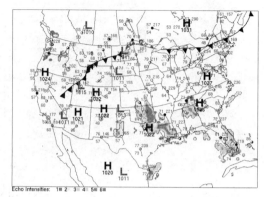

Blackline Master *Climate and Weather* 4.4

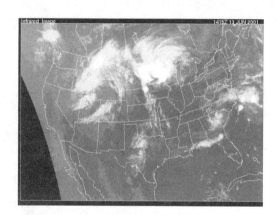 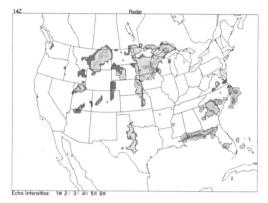

Blackline Master *Climate and Weather* 4.5

 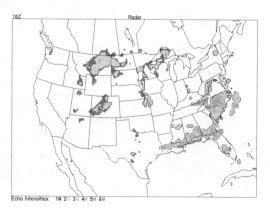

Use with *Climate and Weather* Investigation 4: Radiosondes, Satellites, and Radar

Blackline Master *Climate and Weather* 5.1

The Water Cycle

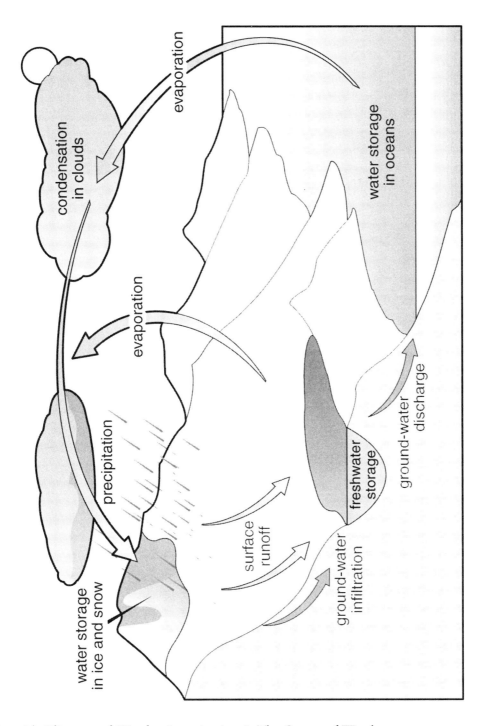

Use with *Climate and Weather* Investigation 5: The Causes of Weather

Blackline Master *Climate and Weather* **5.2**

The Water Cycle

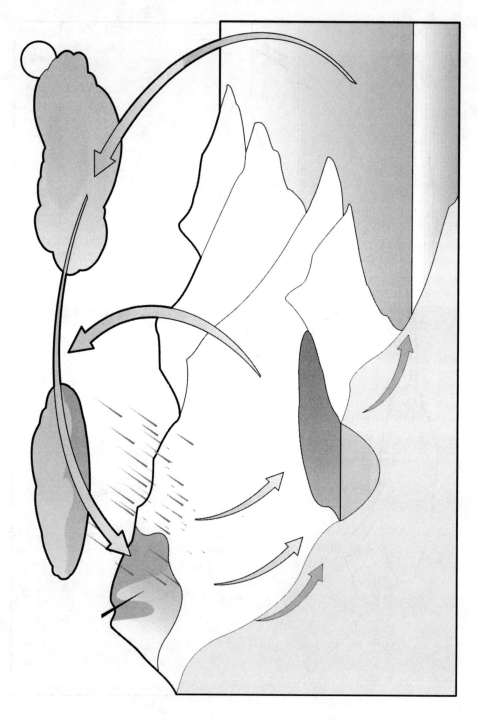

Use with *Climate and Weather* Investigation 5: The Causes of Weather

Blackline Master *Climate and Weather* 6.1

Modified Köppen Classification System: Climatic Regions

Tropical Rain Forest: Over 2.4 in. of precipitation received in all months of the year. Average temperature of coolest month of the year does not fall below 64.4°F.

Tropical Monsoon: At least one month of the year receives less than 2.4 in. of precipitation. Average temperature of coolest month of the year does not fall below 64.4°F.

Tropical Savanna: Summer months are wet, winter months are dry. Average temperature of coolest month of the year does not fall below 64.4°F.

Desert: Dry, evaporation is greater than precipitation by at least half.

Steppe: Semiarid, evaporation is greater than precipitation by less than half.

Mediterranean: Dry summer. Average temperature of coolest month of the year is greater than 32°F, but not more than 64.4°F. Average temperature of warmest month of the year is greater than 50°F.

Humid Subtropical: Some regions experience a dry winter period, while others have no dry season and precipitation exceeds 1.2 in. each month of the year. Temperature of coolest month of the year does not fall below 32°F, and does not rise above 64.4°F. Temperature of warmest month of the year is greater than 50°F.

Marine West Coast: No dry season, precipitation exceeds 1.2 in. each month of the year. Average temperature of coolest month of the year does not fall below 32°F, and does not rise above 64.4°F. Average temperature of warmest month of the year is greater than 50°F.

Humid Continental, Warm Summer: Some regions experience a dry winter period, while others have no dry season and precipitation exceeds 1.2 in. each month of the year. Average temperature of coolest month of the year does not fall below 32°F. Average temperature of warmest month of the year is greater than 50°F. Temperatures vary greatly.

Humid Continental, Cool Summer: Some regions experience a dry winter period, while others have no dry season and precipitation exceeds 1.2 in. each month of the year. Average temperature of coolest month of the year does not fall below 32°F. Average temperature of warmest month of the year is greater than 50°F. Temperatures vary greatly.

Use with *Climate and Weather* Investigation 6: Climates

Blackline Master *Climate and Weather* 6.1 (continued)

Modified Köppen Classification System: Climatic Regions

Subarctic: Some regions experience a dry winter period, while others have no dry season and precipitation exceeds 1.2 in. each month of the year. "Snow" climate. Average temperature of coolest month of the year does not fall below 32°F. Average temperature of warmest month of the year is greater than 50°F. Temperatures vary greatly.

Tundra: Cold, "ice" climate. Average temperature of the warmest month of the year is below 50°F.

Ice cap: Cold, "ice" climate. Average temperature of the warmest month of the year is below 50°F.

Blackline Master *Climate and Weather* 6.2

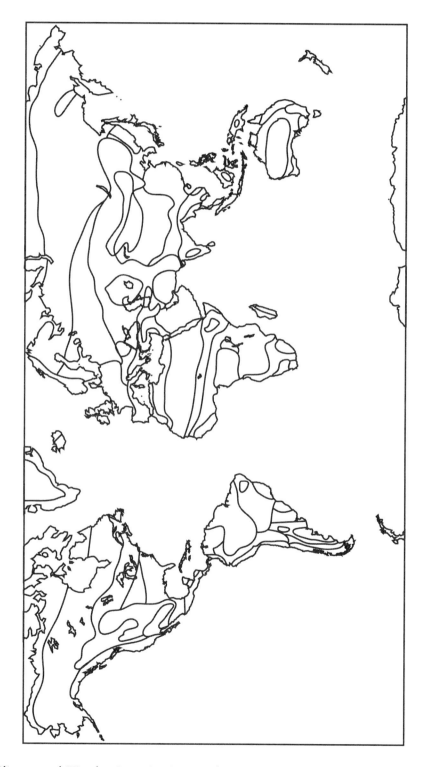

Use with *Climate and Weather* Investigation 6: Climates

Blackline Master *Climate and Weather* 6.3

Temperature Change Data Table			
Temperature	Container #1 (sand)	Container #2 (water)	Container #3 (air)
0 min (starting)			
3 min			
6 min			
9 min			
12 min			

Use with *Climate and Weather* Investigation 6: Climates

NOTES

NOTES

NOTES

NOTES

NOTES

NOTES